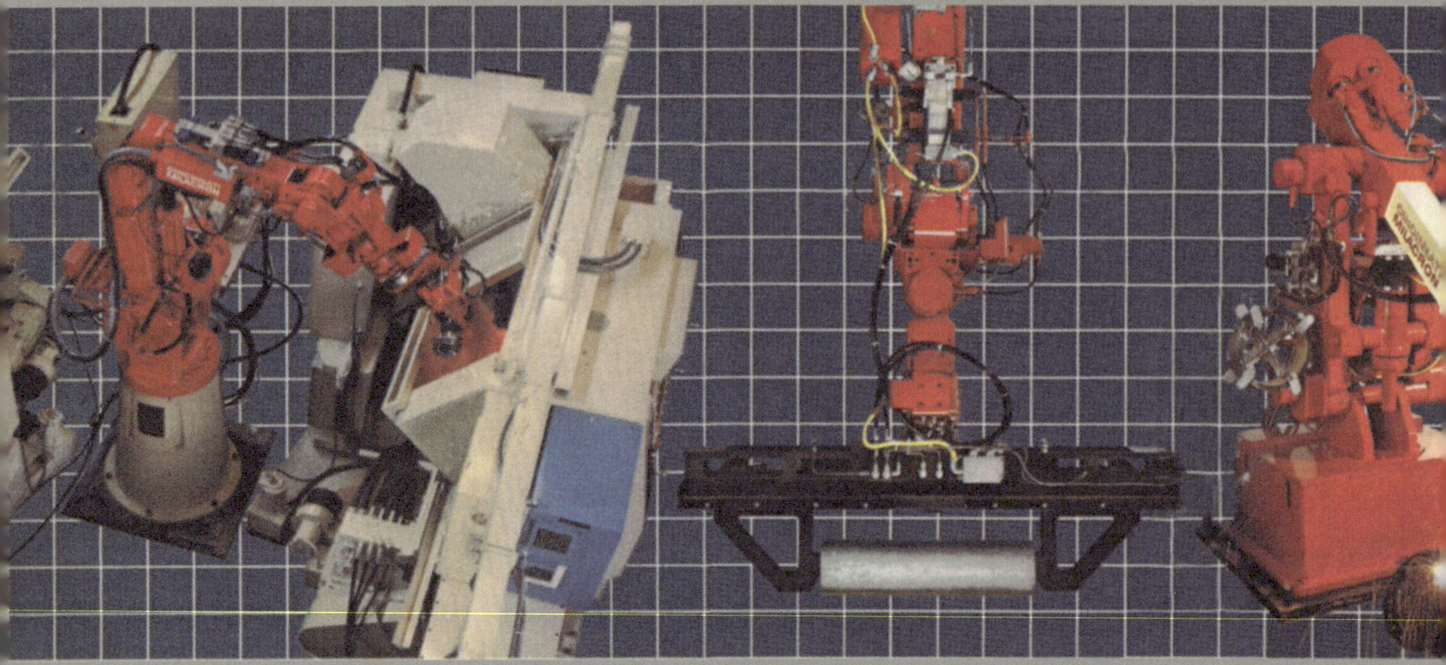

Foreword

THERE are two views about the story of King Canute and the incoming tide. One version has it that he was stupid enough to think that he was such a powerful King that he could stop the tide coming in. The other version suggests that he was demonstrating to his stick-in-the-mud advisers that not even a King as powerful as himself could turn back the tide. Either version has obvious applications to a number of British industrial enterprises.

The simple fact is that there is no way the incoming tide of robots can be turned back. There are three options. Face it out and drown; retreat before it, that roughly equates to importing foreign made robots which means exporting employment; or make the most of it, which entails investing in the rapidly developing robot industry and thereby creating wealth and employment at home.

There is no doubt that the design and manufacturing talent is available. I only hope that the Captains of British Industry will take heed of the second version of the story of King Canute.

Decade of Robotics

Special Tenth Anniversary issue of The Industrial Robot magazine

Contents

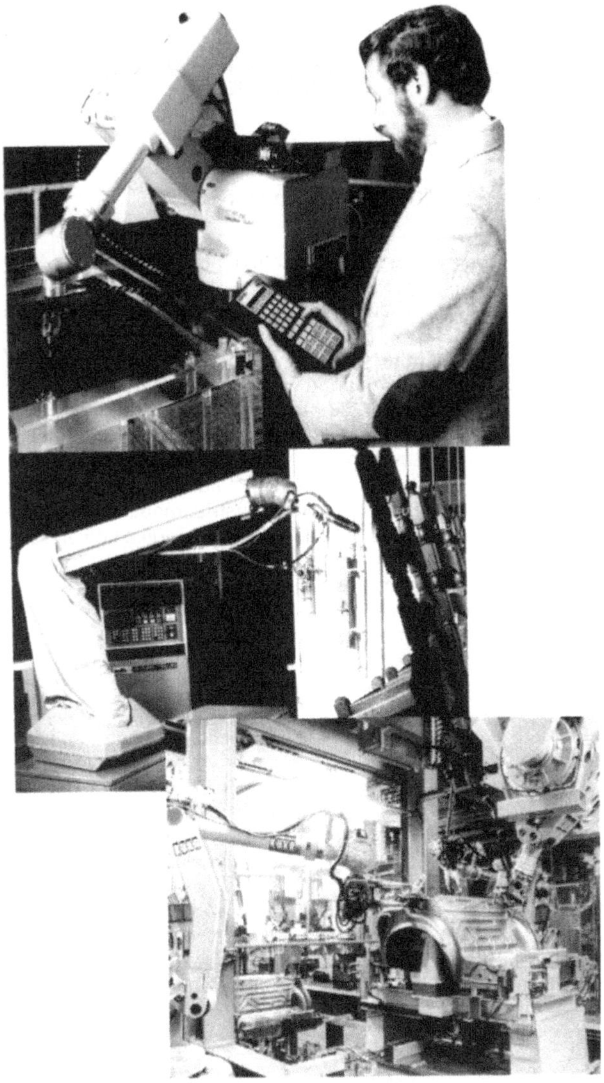

Contents

ISBN 978-3-540-12545-7 ISBN 978-3-662-06501-3 (eBook)
DOI 10.1007/978-3-662-06501-3

'Decade of Robotics' is a stand-alone publication celebrating the Tenth Anniversary of the Industrial Robot magazine and is compiled and edited by the same editorial team. It is co-published by Springer-Verlag, Otto-Suhr-Allee 26/28, D 1000 Berlin 10, GFR and IFS (Publications) Ltd., 35-39 High Street, Kempston, Bedford, UK.
The Industrial Robot magazine is available on subscription at £43 (UK); £53 (elsewhere) per annual volume.

Originally published by IFS(Publications) Ltd., UK. Springer-Verlag · Berlin · Heidelberg · New York · Tokyo in 1983

COVER: A selection of present-day robots. Clockwise: KUKA, Unimate, ASEA, Trallfa, FATA.

Typesetting by Fleetlines Typesetters, Southend-on-Sea and Fotographics (Bedford) Ltd., Bedford.

Printed by The Cotswold Press, Oxford.

The Industrial Robot
Past, present and future

Tom Brock, chairman of IFS, publishers of *The Industrial Robot* magazine answers the question: Why a robot magazine?

LOOKING BACK on a decade of publishing journals in the industrial automation field I find it both surprising and comforting to see that it was only in the second issue of *The Industrial Robot* that we reported the entry of ASEA of Sweden into the robot business. In the same issue we had the privilege of having Joe Engelberger of Unimation Inc. express his opinions in our 'One Man's View' column. These two items are not unconnected.

Engelberger expressed, in no uncertain terms, his strongly held views that 'Unimation Inc., virtually alone, has opted for sophistication; and we diligently seek still further sophistication'. This statement was written in reply to the school of thought that argued for modular construction in the robot business. ASEA's entry into the market place in September 1973 with the first all electric robot certainly supported Engelberger's views. In fact there has not been another robot which has been copied worldwide as much as the ASEA machine introduced in 1973.

'Highly successful worldwide survey'

Who else and what else was there in 1973? For a start there was an opportunity for publishing an industrial robot magazine. IFS recognised this opportunity as a result of its successful first publishing venture in this field in 1972 with a highly successful worldwide survey of industrial robot equipment. In addition, the first truly international conference in Europe on this subject was organised by IFS at Nottingham in March 1973 at which delegates from 15 countries participated.

An international magazine devoted to the emerging technology of robotics was, therefore, an obvious choice for us. The emphasis in *The Industrial Robot* was, and will always remain, both international in coverage and industry-oriented. Readers of *The Industrial Robot* number 8,000 and are located in 42 countries in all 6 continents. The magazine is now accepted as the official publication of two national robot associations.

Today, of course, there are many more publications available dealing with this subject and, whilst we welcome the serious competitors because it is good for the industry to have more than one mouthpiece, we are somewhat concerned by the proliferation of publications in this technology. Will their ability to survive over the next three to five years prove to be better than that of some of the equipment suppliers.

It is interesting to note that of the five companies which advertised in the first issue of *The Industrial Robot* only one is still in the robot business – and that is Trallfa. I believe the robot industry, which achieved worldwide sales in 1982 of only some £250 million, is entering a 'shake-out' period.

The first of the pioneering companies to give in was AMF Versatran in the USA. Not that the equipment or the name disappeared – it is, of course, still available through Prab but AMF decided to call it a day. Another robot pioneer in Europe – Electrolux AB of Sweden – merged its robot interest with those of ASEA. Most recently the founder of the robot industry and standard bearer for so long, Unimation Inc., became part of the Westinghouse Corporation. Other changes can be expected.

These examples clearly illustrate on the one hand the excitement and glamour attached to robots in particular and automated manufacturing in general, and on the other hand the tremendous financial resources required to remain and expand in this business. The industrial robot business is by necessity an international one and in order to compete in world markets robot suppliers will need not only technological expertise but strong sales and marketing abilities which are not usually available to the smaller specialist company.

How will the robot industry develop in the next 10 years? We confidently see a larger and more important role for the computer, both 'on board' machines as well as a process controller and total management tool used in supervising the 1993 factory. We also believe that robot-like devices will be found in many non-industrial applications ranging from undersea exploration and mining to operation in dangerous environments such as fire fighting, window cleaning

on sky scraper buildings to the ever popular 'Martini mixing'. The industrialist will buy his robot as part of a total automation package rather than as stand-alone devices. He may not even be aware that his machines are robot served.

Hopefully we will have learned how to co-operate internationally to ensure an equitable life not only East-West but also North-South. It is to be hoped too that our educational systems will have been overhauled to respond much more rapidly to the changing world in which our children are and will be growing up.

The industrial robot as we know it today has largely been created by one person's dream to remove man from hostile and demeaning jobs. Joe Engelberger was this man – the father of all robots. As we conclude 10 years of publishing *The Industrial Robot* it is a particular pleasure and privilege to record his presence as a midwife at the birth of this publication and his continuing support and interest.

The Industrial Robot has relied on the goodwill and support of many other people worldwide. They are too many to name but I personally would like to thank them all for the co-operation we have enjoyed these past 10 years. I am confident *The Industrial Robot* will be even stronger when it celebrates its second decade.

The Industrial Robot

Passe, present et future

EN FAISANT UNE RÉTROSPECTIVE de dix ans de publications, c'est avec surprise et réconfort que je constate que ce n'est que dans le deuxième numéro de l'*Industrial Robot* seulement que nous reportions l'entrée de l'ASEA, de Suède, dans le domaine des robots. Dans ce même numéro, nous avions le privilège de rapporter les opinions de Joe Engelberger de Unimation Inc., qu'il exprimait dans notre rubrique "point de vue d'un homme". Ces deux faits ne sont pas sans relation. M. Engelberger exprima, sans equivoque, ses idées très fermes "qu'Unimation Inc., pratiquement seule, a opté pour la sophistication et nous recherchons avec assiduité une plus grande sophistication." Cette déclaration était écrite en réponse à l'école de pensée qui favorisait la construction modulaire dans le domaine des robots. L'entrée sur le marché, en septembre 1973, de l'ASEA avec le premier robot entièrement électrique supportait certainement le point de vue d'Engelberger. En fait, aucun autre robot n'a autant été copié dans le monde que la machine d'ASEA introduite en 1973.

Qui et quoi d'autre en 1973? Au départ, une opportunité éxistait pour la publication d'un magazine traitant des robots industriels. IFS prit conscience de cette opportunité grâce à sa première tentative, couronnée de success, en matière de publication dans ce domaine en 1972, sous la forme d'une enquête mondiale, réussie, sur les équipements robotiques industriels. De plus, la première conférence de caractère réellement international en Europe, consacrée à ce sujet fut organisée à Nottingham en 1973 par IFS. Des délégués representant 15 pays y ont participé. Un magazine international consacré à la technologie naissante de la robotique était donc pour nous un choix évident. Le trait principal de l'*Industrial Robot* a été et restera toujours d'une part lié à l'information internationale et d'autre part orienté vers l'industrie. Les lecteurs de l'*Industrial Robot* sont au nombre de 8,000 dans 42 pays sur les six continents. Le magazine est maintenant reconnu comme étant la publication officielle de deux associations nationales de robotique.

Aujourd'hui il y a bien sur beaucoup d'autres publications qui traitent de ce sujet et bien que nous réjouissions de la venue de concurrents sérieux, parce qu'il est bon pour l'industrie d'avoir plus d'un porte-parole nous sommes néanmoins inquiets de la prolifération des publications dans le domaine de cette technologie. Leur chance de survie dans les trois à cinq années à venir sera-t-elle meilleure que celle de certains fournisseurs de l'équipement?

Il est intéresssant de remarquer que parmi les cinq compagnies qui faisaient de la publicité dans le premier numéro de l'*Industrial Robot* une seule a survécu dans le domaine des robots et il s'agit de Trallfa. J'ai le sentiment que l'industrie des robots qui en 1982 a atteint un chiffre de ventes de 250 millions de £ seulement, entre dans une période de "bouleversement".

Parmi les compagnies d'avant-garde dans ce domaine, la première à avoir abandonné fut AMF Versatran aux Etats-Unis. Non pas que les équipements et le nom aient disparu, ils sont toujours disponibles par l'intermédiaire de Prab, mais AMF décida de laisser tomber. Un autre pionnier des robots en Europe – Electrolux AB, de Suède, a fusionné ses intérêts en matière de robot avec ceux de l'ASEA. Plus récemment, le fondateur de l'industrie du robot et son porte-drapeau pendant si longtemps, Unimation Inc. a été rattaché à la Westinghouse Corporation. On peut s'attendre à d'autres changements.

Ces exemples montrent clairement d'une part l'éxaltation et le prestige liés aux robots en particulier et à la fabrication automatisée en general, et d'autre part les énormes ressources financières nécessaires à la survie et à l'expansion dans ce domaine. Le commerce des robots industriels est par nécessité un commerce international et les fournisseurs de robots, s'ils veulent être compétitifs sur les marchés mondiaux, devront non seulement avoir l'expertise technologique mais aussi un niveau de ventes élevé ainsi qu'une compétence en marketing dont les petites compagnies spécialisées ne disposent pas généralement.

Comment l'industrie des robots se developpera-t-elle dans les dix prochaines années? Nous voyons avec confiance l'ordinateur jouer un rôle plus grand et plus important dans la surveillance de l'usine de 1993, qu'il soit "a bord" des robots, contrôleur de procédé ou outil de gestion totale. Nous pensons également que les appareils de type robot existeront dans beaucoup d'applications non-industrielles allant de l'exploration et du forage sous-marins aux opérations dans des situations dangereuses telles que la lutte contre le feu, le nettoyage des vitres des grattes-ciel en passant par le "Martini Mixer" toujours populaire. L'industriel achêtera son robot comme l'élément d'un ensemble d'automatisation totale plutôt que comme des appareils individuels. Il pourrait même ne pas réaliser que ses machines sont actionées par des robots.

Il faut espérer que nous aurons appris à coopérer au niveau international pour assurer une vie équitable non seulement de l'Est à l'Ouest mais aussis du Nord au Sud. Espérons également que nos systèmes d'éducation auront été revus pour répondre de façon plus rapide aux besoins d'un monde en mutation dans lequel nos enfants sont, et seront, en train de grandir.

Le robot industriel tel que nous le connaissons aujourd'hui a été largement créé par le rêve d'un homme d'éloigner l'homme des travaux hostiles et dégradants. Cet homme était Joe Engelberger, le père de tous les robots. Alors que nous achevons dix ans de publication de l'*Industrial Robot* nous avons le plaisir et le privilège tout particulier de noter sa présence, telle une sage-femme, à la naissance de cette publication et de noter son intérêt et son soutien constants. L'*Industrial Robot* a compté sur la bonne volonté et le soutien de beaucoup d'autres personnes dans le monde entier. Ils sont trop nombreux pour être nommées, mais j'aimerais personnellement les remercier tous pour leur co-operation pendant ces dix dernières années. Je crois fermement que l'*Industrial Robot* aura encore plus de portée quand il célèbrera sa deuxième décennie.

The Industrial Robot

Vergangenheit, Gegenwart und Zukunft

WENN wir auf ein Jahrzehnt, in dem wir Zeitschriften aus dem industriellen Automationsgebeit veröffentlicht haben, zurückblicken, finde ich es sowohl erstaunlich als auch tröstlich, daß wir schon in der zweiten Ausgabe des *Industrieroboters* von dem Eintritt der ASEA aus Schweden in das Robotergeschäft berichtet haben. In derselben Ausgabe hatten wir das Privileg, daß Joe Engelberger von Unimation Inc. seine Ansichten in unserer Spalte "Ansichten eines Mannes" vertrat. Diese zwei Artikel stehen im Zusammenhang miteinander. Engelberger drückte mit bestimmter Klarheit seine festen Ansichten darüber aus, daß Unimation Inc. sich im Grunde genommen ausschließlich für die Raffinesse entschieden hat; und daß wir weiterhin fleißig nach Raffinesse suchen. Diese Aussage wurde als Antwort auf die Denkschulrichtung geschrieben, die sich für ein Baukastensystem im Robotergeschäft einsetzte. Der Eintritt der ASEA in den Markt im September 1973 mit dem ersten vollelektrischen Roboter unterstützte sicherlich Engelbergers Ansichten. Tatsache ist, daß es keinen anderen Roboter gegeben hat, der in der ganzen Welt so viel nachgemacht wurde wie die ASEA Maschine, die 1973 eingeführt wurde.

Wen und was anderes gab es 1973? Zunächst gab es Gelegenheit zur Veröffentlichung. IFS erkannte diese Gelegenheit aus dem Ergebnis seines ersten erfolgreichen Veröffentlichungsunternehmens in diesem Gebiet im Jahre 1972, das einen höchst erfolgreichen weltweiten Bericht über industrielles Robotermaterial beinhaltete. Außerdem wurde die erste echte internationale Konferenz in Europa mit diesem Thema von der IFS in Nottingham im März 1973 organisiert. Es nahmen Vertreter von fünfzehn Ländern teil. Aus deisem Grunde war die Entscheidung für eine internationale Zeitschrift, die sich der entwickelnden Technik von Robotistik widmete, klar. Der Schwerpunkt in dem *Industrieroboter* war und wird immer sowohl international in seinem Geltungsbereich als auch industriebezogen bleiben. Die Leser des *Industrieroboters* (8000) befinden sich in 42 Ländern in allen sechs Kontinenten. Die Zeitschrift ist jetzt von zwei nationalen Robotervereinigungen als eine offizielle Veröffentlichung akzeptiert worden.

Heute stehen natürlich viel mehr Veröffentlichungen zur Verfügung, und wenn wir auch die ernstzunehmenden Konkurrenten begrüßen, weil es für die Industrie gut ist, mehr als ein Sprachrohr zu haben, so beunruhigt uns doch irgendwie die Vermehrung der Veröffentlichungen in dieser Technik. Wird ihre Überlebungsfähigkreit über die nächsten drei bis fünf Jahre sich als besser erweisen als die einiger Materiallieferanten?

Es ist interessant zu bemerken, daß von den fünf Gesellschaften, die in der ersten Ausgabe des *Industrieroboters* annoncierten, nur noch eine, nämlich Trallfa, im Robotergeschäft ist. Ich glaube, daß die Roboterindustrie, die 1982 einen weltweiten Umsatz von nur 250 Millionen erzielte, in eine Periode der Auslese eingetreten ist.

Die erste der bahnbrechenden Gesellschaften die aufgab, war AMF Versatran in den USA. Es war nicht so, daß dad Material oder der Name verschwand – das steht natürlich noch zur Verfügung durch Prab, aber AMF entschloß sich, zu schließen. Ein anderer Roboterbahnbrecher in Europa – Electrolux AB aus Schweden legte seine Roboterinteressen mit denen von ASEA zusammen. Vor kurzem wurde der Gründer der Roboterindustrie und Wegweiser für lange Zeit, Unimation Inc., Teil der Westinghouse Corporation. Andere Veränderungen sind zu erwarten.

Diese Beispiele zeigen auf der einen Seite deutlich das Interesse und den Glanz, mit denen die Roboter im besonderen und die automatisierte Produktion im allgemeinen umgeben sind; auf der anderen Seite weisen sie auf die ungeheueren finanziellen Mittel hin, die erfgorderlich sind, um im Geschäft zu bleiben, und um es auszudehnen. Das industrielle Robotergeschäft ist notwendigerweise ein internationales Geschäft und, um mit dem Weltmarkt zu konkurrieren, benötigen die Roboterlieferanten nicht nur technische Fachkenntnis, sondern auch starke Absatz- und Marketing- fähigkeiten, die gewöhnlich nicht den kleinen Spezialistengesellschaften zur Verfügung stehen.

Wie wird sich die Roboterindustrie in den nächsten zehn Jahren entwickeln? Wir sind davon überzeugt, daß der Computer eine größere und bedeutendere Rolle in der Überwachung der Fabrik von 1993 spielen wird, sowohl dezentral an den Robotern, als auch als Prozeßsteuerung und als Gesamtbetriebssystem. Wir glauben auch, daß roboterartige Einrichtungen in vielen nicht-industriellen Anwendungsbereichen gefunden werden können. Diese Gebiete reichen von Unterwasserforschung und Bergbau, vom Einsatz in gefährlichen Bereichen, wie Kampf mit dem Feuer, Fensterputzen an Wolkenkratzerngebäuden hin bis zu dem berühmten "Martinimixer". Der Industrialist wird seinen Roboter eher als Teil eines ganzen Automationspackets kaufen, als ihn als isolierte Einrichtung erwerben. Er ist sich vielleicht noch nicht einmal bewußt, daß seine Maschinen von Robotern bedient werden.

Wir hoffen, daß wir gelernt haben, international zusammenzuarbeiten, damit ein gerechtes Leben nicht nur im Osten – Westen sondern auch im Norden – Süden gesichert ist. Wir hoffen auch, daß unser Erziehungssystem gründlich geprüft wird, damit es sich viel schneller der verändernden Welt anpassen kann, in der unsere Kinder jetzt leben und aufwachsen werden.

Der Industrieroboter ist, wie wir ihn heute kennen, zum größten Teil von dem Traum eines Mannes, den Menschen von feindlichen und erniedrigenden Arbeiten fernzuhalten, geschaffen worden. Joe Engelberger war dieser Mann-der Vater aller Roboter. Da wir zehn Jahre der Veröffentlichung des *Industrieroboters* abschließen, ist es uns eine besondere Freude und ein besonderes Privileg, seine Anwesenheit als Hebamme bei der Geburt dieser Veröffentlichung und seine fortwährende Unterstützung und sein Interesse aufzuzeichnen. Der *Industrieroboter* hat auch auf den guten Willen und die Unterstützung von vielen anderen Menschen weltweit gebaut. Es wären zu viele, um sie alle zu nennen, aber ich möchte mich persönlich bei allen für die Zusammenarbeit in den vergangenen zehn Jahren bedanken. Ich bin sicher, daß der *Industrieroboter* noch bedeutender sein wird, wenn er das zweite Jahrzehnt feiert.

The cocktail party that gave birth to the robot

Dr Brian Rooks, Editor of *The Industrial Robot*, traces the birth and evolution of the robot

ALTHOUGH the word 'robot' is a 20th century creation the concept of the human-like machine goes back much further. In the 18th century Swiss craftsmen built a number of life-like automotans able to write, draw and play musical instruments.

Writers have often dwelt on the fear of the 'machine monster' created by man in his own image – *Frankenstein* by Mary Shelley published in 1817 was one of the first works on the subject. Since then writers, movie makers and television producers have exploited every facet of humanoid machines, or robots.

In general the emphasis has been on the danger of robot 'take-overs'. This was the theme of the play by the Czechoslovakian, Karel Capek, in which the word 'robot' was first used; robot being derived from the Czech word for 'worker'. The theme of the play, entitled R.U.R. or Rossums Universal Robots, is the creation by Rossum of a line of robots that are meant to relieve mankind of work. The play ends by the perfect robots becoming intolerant of the imperfect humans and exterminating all human life.

'Not all writers are paranoid'

However, not all writers have been paranoid of the all-powerful, all-destructive image of the robot. Isaac Asimov was able to create another image, namely that of the benevolent robot. He created the three laws of robotics which have become standards that all aspiring robot designers and users should attempt to meet. The first and most important was that 'a robot must not harm a human being nor through inaction allow one to come to harm.'

In spite of the writings of Asimov and of the efforts of the early pioneers robots have had to carry 'the cross' of repulsiveness that is associated with the word. To this day many feel that the choice of this word has had a detrimental effect on the application of the programmable machines that are classified under the title of industrial robots. Fortunately such minor details did not deter the men at the birth of the industrial robot and now some 20 or so years later these machines are beginning to be used in significant numbers.

The real birth of the industrial robot can be traced accurately to a meeting in 1956 at a cocktail party between George C. Devol and Joseph Engleberger, as the latter recounts in the preface to his book *Robots in Practice.* Devol, an inventor with a lineage of patents in the electronics field, had applied for a patent labelled *Programmed Transfer Article.* To Engleberger, the entrepreneur and avid reader of Asimov this introduction proved to be the spur that fired him on a life-long journey of robotics that has earned him the name the 'Father of Robotics'. Devol's patent was issued in 1961 as US Patent 2,988,237. This was the basis of the Unimate robot that first saw the light of day in 1960.

Devol's was not the first patent on robotics. In fact, it was pre-empted by one from a British inventor, Cyril Walter Kenward, issued in 1957 as British Patent 781,465. Unimation and another US robot builder AMF infringed this patent and ultimately a cash settlement was made to Kenward. If only Kenward had been able to meet up with a British 'Joe Engelberger' then the story of industrial robot development might have been so different.

As it was, Unimates began to be made in the early days as hand built machines. The first one was sold in 1961 to the Ford Motor Corporation to tend a die casting machine. It is perhaps ironic that the first robot went to a company who refused to recognise it as a robot because the word 'robot' conjured up the thought of 'a destroyer of jobs'. Ford adopted the title UTD (universal transfer device) and even to this day that is the preferred name in Ford. That first robot was withdrawn from service a few years ago and is now a museum piece at the Smithsonian Institute in Washington, DC.

Initially, Unimates and its rival, the Versatran from AMF, were seen as substitutes for manual operators in dull, repetitive jobs such serving die casting machines, injection moulding machines, presses, metal cutting machine tools and the like. The emphasis was on manipulating parts and not tools.

It was a small Norwegian agricultural company that started the concept of

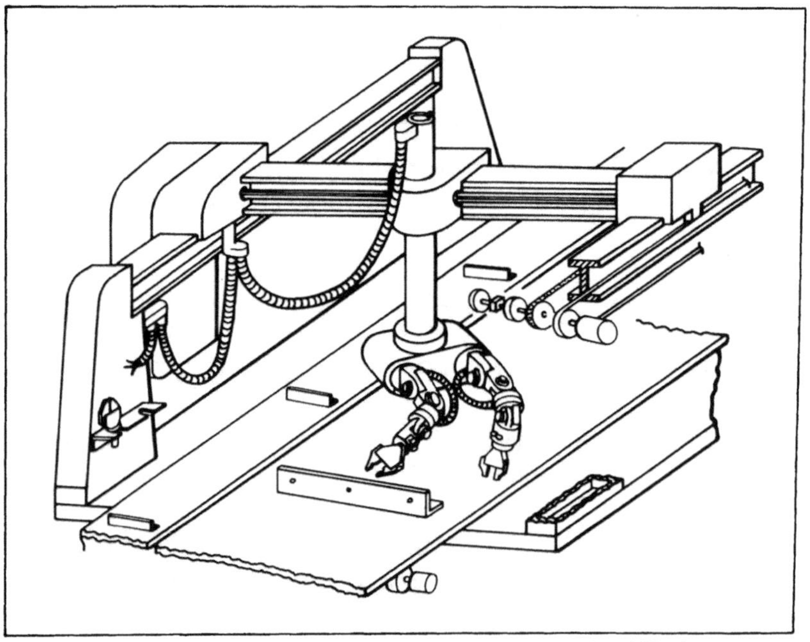

Fig. 1. British Patent 781,465 dated March 29 1954 refers to the first UK robot.

using the robot to manipulate a processing tool. The Trallfa painting robot was born out of the necessity to solve a problem. This was the opposite to the case of the Unimate where the solution preceded the problem.

In 1964 the painting of wheelbarrows at Trallfa's factory in Byrne was becoming a problem because of the difficulty of recruiting workers from a predominantly agricultural district to do a dirty job. Fortunately within the company was an extremely talented engineer, Ole Molaug who was set to work on producing an automatic painting machine.

By 1966 the first prototype painting robots had been installed in the Byrne plant to coat wheelbarrows. By 1969 the Trallfa had sold the robot to a customer. Two machines were installed at Gustavsberg in Sweden for the enamelling of sanitary ware, such as bath tubs, shower trays and the like. These are still working.

Painting was closely followed by spot welding. It turned out that the Unimate was an ideal spot welding automat and in 1968 Unimation took its first multirobot order from General Motors. These robots, 26 in all, were installed in 1969 at GM's Lordstown plant to assemble Vega car bodies. In 1972 the first spot welding line in Europe went into action at Fiat in Italy. That year also saw Unimation's Japanese licencee, Kawasaki, put a line in at Nissan Motors. So started the robot revolution in Japan.

By the 1970s related events had begun to occur that indicated the subject was becoming more accepted. The first symposium on industrial robots took place in Chicago in 1970. This was a US national event organised by IITRII but the next year it went international. Again this was arranged by IITRI and took place in Chicago as the second International Symposium on Industrial Robots. It has been held every year ever since, rotating in the sequence USA-Europe-Japan-Europe. This year it has reached number thirteen and returns to Chicago for the first time since 1975.

Japan adopted the robot with enthusiasm and in 1971 was the first nation to form a robot association – the Japan Industrial Robot Association (JIRA). It was not until 1975 that the Robot Institute of America (RIA) was formed in the USA. The 'Brits' came in with the British Robot Association or BRA in 1977 and since then many countries have formed national bodies, some as trade associations like JIRA, others an amalgam of manufacturers and users like RIA and BRA. In addition to organisations in industrial countries like France, Italy and Sweden associations have been formed in Denmark, Belgium, Australia and Singapore.

In 1973 IFS launched the first international journal devoted to the subject *The Industrial Robot.* Although JIRA also published a journal *(Robot)* in the same year it was in the Japanese language and covered only the domestic scene. In the last year or so *The Industrial Robot* has been joined by one or two other English language journals published in the USA.

Fig. 2. Joe Engelberger with a Unimate.

'Unimates set the scene and were copied'

It is often said that the Unimates being built today are no different to the first machines produced in the early 1960s. In many respects this is true mechanically, but significant advances have been made on controllers and in software. It is true that the hydraulically operated Unimates set the scene and were copied in the early days. But even the number of Unimate copies is small compared with the number of ASEA IRb lookalikes that now adorn the world's industries.

ASEA introduced its all-electric drive IRb6 robot in 1973 and the IRb 60 in 1974. It is often thought that these machines were the first electric robots but this is not strictly true. As a survey published by IFS in 1972 showed out of 133 listed robots there were six electric robots. All but one of these, including one from Sundstrand, the US machine tool company, have since disappeared. The one remaining is Yaskawa which in 1972 had a robot called Motohand. Its current range is sold as Motoman and has more than a passing resemblance to ASEA's robots. It can be said in all fairness that the ASEA IRb 6 was the first commercial all-electric drive robot.

The all-electric robot now dominates the arc welding field but the application started before such machines became established. The claim to the first use of a robot for this operation is often claimed by Kawasaki when, in 1974, it installed a Kawasaki-Unimate to weld motor cycle frames.

In fact this was preceded by a development at Hawker Siddeley which in the late 1960s and early 1970s was building AMF Versatrans under licence in the UK. In 1972 the UK company had adapted a Versatran to carry out welding and gave it the name Versaweld. This machine was developed for British Rail trucks. It had another innovation at that time – it was mounted on tracks to give an additional traversing axis.

'Big expansion in applications'

The first generation robot encompassed mid-1950s NC machine tool technology of hard-wired controllers and either wire or pegboard memories. By the 1960s NC controllers were moving to CNC technology. Therefore it was no surprise that when the first machine tool company entered the robot business it introduced this technology to robots.

Cincinnati Milacron was that company. In 1975 it announced on the open market its computer-controlled 6CH robot, later re-titled the **T**^{3}.

In the latter half of the 1970s there was a big expansion in the range of robot applications attempted by companies. One growth was in the use of the robot as a manipulator of metal removal tools, mainly through the efforts of Cincinnati and ASEA.

Cincinnati's pioneering work was mainly in using the robot for drilling, tapping, milling operations whereas ASEA's initiatives were into the areas of grinding, including such operations as fettling, deburring and polishing.

The impetus for using the robot as a drilling/tapping machine came from the United States aerospace industry. Cincinnati were able to enter into this development partly through their experience as a machine tool supplier to the aerospace industry of their T^3 robot. The main applications have been the drilling and routing of airframe panels.

Drilling and the like tend to be restricted to the aerospace industries. On the other hand the pioneering work of ASEA in the grinding field is over a much wider range of general engineering industries. These include automotive, domestic appliance, food as well as in foundries.

The first known example of using a robot for fettling in an iron foundry was at the works of Kockums at its Kallinge foundry in southern Sweden. This was installed in 1976. The machine being used was an ASEA 60kg robot fitted with 5 hp air motor. The grinding wheel is forced against the casting by pneumatic pressure so that a constant force was maintained between the wheel and the casting. The castings are manually loaded into a simple jig at one station of a two station rotary table. At the other station the robot grinds the casting.

Most experts agree that the biggest potential for robots is in assembly. There are many problems to be solved, particularly in parts feeding and presentation. Meanwhile a start has been made on developing suitable robots.

Compared with earlier robots designed for relatively heavy duty operations an assembly robot has to be fast and precise. To meet these requirements the Italian company Olivetti moved away from the conventional single arm floor-mounted robot and produced a multi-armed gantry machine. This was produced in 1975 as the *Sigma*. In the main, the Olivetti robots have been used for electronic component insertion and typewriter keyboard assembly.

One of the most well known robots initially designed for assembly is the PUMA (Programmable Universal Machine for Assembly) which was first seen in 1978. PUMA's are manufactured by Unimation and were

Fig. 3. Hall Automation RAMP spray painting robot.

developed to meet a General Motors specification for a robot with a space intrusion comparable to a human's and a weight-handling capacity of 5 lbs. Several hundred of these machines have now been sold although not all are in assembly. They have been applied to handling, arc welding, sealant application and others.

Since then other assembly robots have appeared, notably the PRAGMA from DEA in 1979, which is now licensed to General Electric in the USA, and the SCARA robot developed at Yamanashi University in Japan. SCARA, an acronym for Selective Compliance Assembly Robot Arm, is one of the few innovative robot ideas to emerge from Japan. The selective compliance stems from the fact that it is extremely rigid in the vertical direction but compliant horizontally. The former allows the robot to press, punch, drill and perform other assembly process operations in that direction. The degree of compliance in the latter is sufficient for 'peg-in-the-hole' type operations.

The SCARA concept has been adopted by a number of Japanese companies which supported the development at Yamanashi. Commercial machines were introduced in 1981 by Sankyo Seiki, Pentel, Nitto Seiko and NEC. The Sankyo Scara is now marketed by IBM as the 7535 robot.

Supposedly the 'last nut to crack' in robotics is to provide machines with intelligence which in the main means giving them senses of sight and touch. Work started on vision in the 1960s by artificial intelligence teams at SRI in the USA and Edinburgh University in the UK.

Their concentration was on solving general problems of pattern recognition rather than specific industrial jobs. Later SRI turned its attention to the latter as did Nottingham University. The Nottingham team demonstrated in 1972 its SIRCH machine which was capable of being taught to recognise two-dimensional parts and to subsequently identify and orientate those parts when presented in a random non-orientated manner. However it was not until the 1980s that such techniques were even considered for industrial application. There were many reasons for this delay, including industry's inability to grasp the basic techniques involved, the difficulty of 'talking' with robot controllers of the 1970s and the justification.

General Motors was probably the first company to apply vision in an industrial situation. Following development work based on SRI's work at the GM Tech Centre its Consight system was installed at a foundry in St. Catherines, Ontario, Canada. Castings passing along a conveyor belt from the shake-out area are recognised and placed by a T^3 robot into the appropriate bin. Renault in France also has been active in this area and in 1982 used its own vision system to guide a Renault V80 robot to pick crankshafts from a palletised stack and place them into a grinding machine.

Fig. 4. Cincinnati robot making parts for robots.

'Force sensing – the biggest advance'

The less glamorous subject of touch and force sensing has paralleled the vision activities. In 1974 Hitachi demonstrated its Hi-T-Hand robot which employed force feed-back in the robot hand to guide pins into holes and similar operations. The Olivetti SIGMA machine also incorporates force and touch sensing in the grippers, as does the DEA PRAGMA.

The biggest advance in this field took place in 1976 when, following research at the Charles Stark Draper Laboratories in Cambridge, USA into the dynamics of part mating, the Remote Center Compliance (RCC) was evolved. This is a passive device with no feed-back, but which allows the device to distort as a shaft, for instance, is pushed towards a hole allowing the peg to line up laterally and rotationally with hole. Thus a peg can be inserted into a hole with only a few microns clearance. Commercial versions are now available.

The biggest changes now taking place in robotics are in the areas of control and software. Languages are being developed which should ease programming for complex operations. Off-line programming and voice communication are the most likely developments that will advance robotics in the next five years. However the major problems of application engineering and economic justification will continue to dominate the subject for several years to come.

The household robot: By 1993

Joe Engelberger, head of Unimation Inc., looks forward to the day when robots man the home

ONCE some years ago Nobelist Dennis Gabor explained to me that being a futurist was jolly good fun as well as a reward for having gained some notoriety in any arena of one's choice. He offered one caveat, however. 'Always make your predictions for events at least 50 years in the future. This will ensure that you are not going to be around to stand muster when your crystal ball is shown to be cloudy.'

Well, my arena of choice is robotics and my assignment for this 10th Anniversary Edition of The Industrial Robot is the future. But, Tom and Brian would not let me off with a good solid forecast for the robotics situation in 2033. My assignment is the immediate 10 years ahead! Forgive me, Dennis, for not heeding your warning.

A gentle jumping off point might be a list of jobs that robots are now doing. Fig. 1 may not be exhaustive, but it does record jobs that robots do successfully and with proven economic justification. While the first industrial robot was installed 22 years ago, by 1983 none of these applications had come close to being saturated.

So, the first prediction for the next decade is that robots will continue to penetrate these job markets. One may hope that misguided educators will not persist in training humans for such vanishing industrial roles.

In Fig. 2 there is an even less exhaustive list of jobs for which we can be reasonably sure that robots will handle successfully and economically within the next five years. By the time of this Anniversary Issue publication one or more may already have earned a transfer to the first list.

In general these jobs cannot be done simply by installing robots built to currently available commercial technology. There is need for adding superior qualities such as adaptability, dexterity and rudimentary perception to the robot or perhaps to its surroundings. Much of what can be contributed to robotics in the decade ahead will be done in peripherals.

I have said that the Fig. 2 list is not exhaustive. Of that I am quite certain because robotics has proven to be a particularly serendipitous activity. The pioneers who developed the first robot never thought of spot welding, the largest single robot activity. Investment casting is another surprise application with the robot proving to be so spectacularly adept that the very activity is enjoying strong growth.

'Gee whiz, when will I get something meaty?'

In Unimation Inc. we are ever being delighted by the ideas put forth by prospective customers who rightly still look upon the robot as a 'solution looking for problems.'

Fig. 3 is the 1993 list and, of course, it is the least disciplined and least exhaustive of all. However, it does represent such a spectrum of activities as to defy any effort to circumscribe robotics ever again. By 1993 it will be evident that the work-a-day world is the robot's oyster.

The activities listed in Fig. 3 were not just extracted from a fertile imagination. There are experiments going on with robots in each of these diverse applications and preliminary results are already encouraging. To be sure innovation and perhaps even invention will be necessary before some of these jobs are done successfully and on a sound economic basis. Still, there is lots of time between now and 1993 and robotics has finally achieved a self-nurturing momentum.

To justify the conviction regarding such a plethora of new robot applications, I call upon the results of a 1968 study by the US Air Force. It was called Project Hindsight.

The objective of the study was to determine the critical factors essential to success of any innovation. The conclusion was that innovations succeed only if all of the following factors are present:

☐ A perceived need.
☐ Availability of appropriate technology and competent practitioners.
☐ Adequate financial support.

It was not everso; but, for the decade ahead, all of the criteria will be in place for dramatic advances in robotic applications.

Some of the bits and pieces that are still not in hand are worthy of comment. Robots need rudimentary vision, tactile sense, mobility, voice communication and inherent safety.

For all of these attributes a level of availability already exists and the relevant arts are being supplemented on a continuing basis in academic and industrial laboratories around the world.

It seems today to be more likely that capability will be enhanced throughout the decade rather than be benefited by quantum leaps of invention. In any event I predict that all we shall need for the roles forecasted will be in place before the end of The Industrial Robot's next decade.

By way of illustration Fig. 4, 5 and 6 show three current embodiments of robot vision, tactile sense and mobility. Fig. 4 is a vision system that is mounted

CURRENT APPLICATIONS

Die Casting	Machine Loading
Spot Welding	Stamping
Arc Welding	Plastic Molding
Glass Handling	Investment Casting
Heat Treatment	Conveyor Transfer
Forging	Palletizing
Paint Spraying	Inspection

Fig. 1. Jobs robot now do.

SOME NEAR TERM APPLICATIONS

Batch Assembly
Order Picking
Wire Harness Manufacturing
Packaging
Textiles Processing
Medical Lab Handling
Fettling

Fig. 2. Jobs robots can expect to do.

FARTHER OUT APPLICATIONS

Garbage Collection
Fast Food Preparation and Delivery
Gasoline Dispensing
Animal Husbandry
Nuclear Maintenance and Cleanup
Hospital Aides
Prosthesis
Neurosurgery
Household Servant

Fig. 3. 1993 – the work-a-day world.

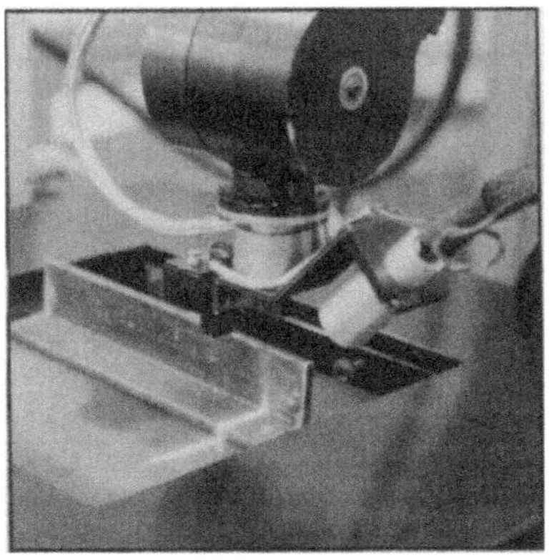

Fig. 4. Vision in hand.

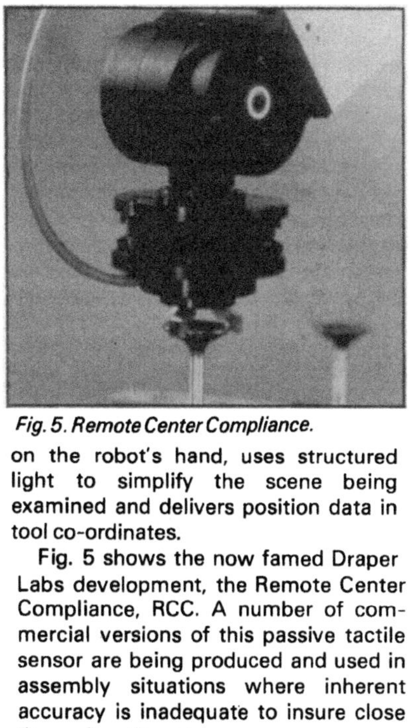

Fig. 5. Remote Center Compliance.

on the robot's hand, uses structured light to simplify the scene being examined and delivers position data in tool co-ordinates.

Fig. 5 shows the now famed Draper Labs development, the Remote Center Compliance, RCC. A number of commercial versions of this passive tactile sensor are being produced and used in assembly situations where inherent accuracy is inadequate to insure close fit up of mating parts.

Fig. 6 depicts a robot arm on a computer controlled cart. The unique mobile robot development, owing in part to Stanford University graduate students, can direct the robot arm to destinations with an accuracy of 1% of distanced travelled. It then uses docking instrumentation to find its precise commanded location.

And so it goes on. Voice communication will be a gift to robotics from the Office-of-the-Future. Going from voice

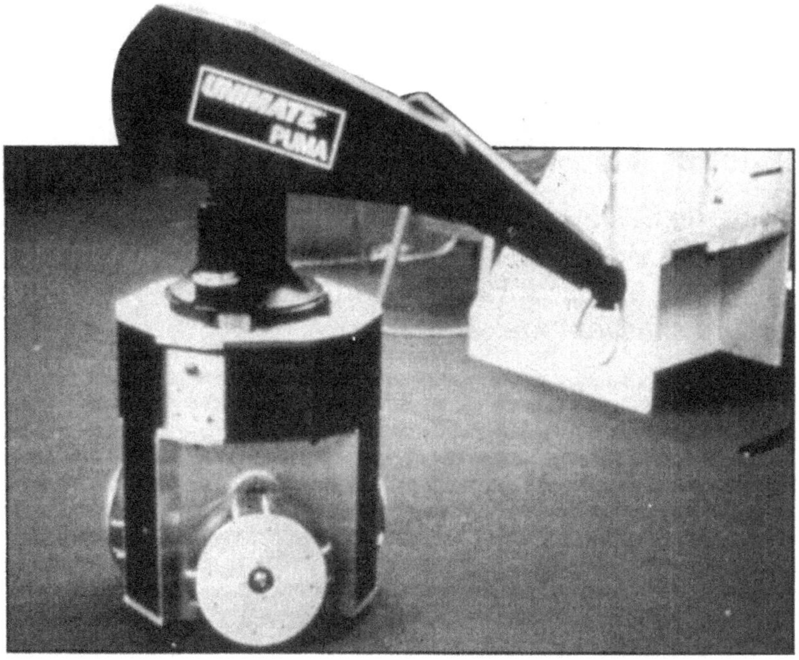

Fig. 6. Robots on wheels.

to typewritten documents is such a heavily 'perceived need' that it is surely getting more than 'adequate financial support.' We roboticists will climb on the backs of the IBMs and Xeroxs and so on of this world.

Inherent safety was Asimov's first requirement. 'A robot may not harm a human nor through inaction allow a human to come to harm.' Having already logged the first fatality at the hand of a robot, the industry worldwide is concerning itself with safety – at least with regard to robot action.

Conjecture about the next 10 years would not be complete if it did not recognise that robots do not stand alone. There are powerful related technologies such as CAD-CAM and Group Technology that will build a friendly environment for robots in factories. And robots will be a key element in the 'Factory of the Future' or, if you will, using Japanese terminology, in Methodology for Unmanned Manufacture, MUM.

Gee whiz, when will I ever get to something meaty for businessmen, like a projection of market growth? Not at all, I am sorry, even though I would be baffled by an average growth rate of less than 35% annually through the decade.

Let me share instead a thought owing to Unimation Inc's consultant, Arthur D. Little. They advised us to forget estimating market size and to instead assume an infinite sump. The problem is not market size, it is rate of conversion.

The jobs listed in Fig. 1, 2 and 3 offer an essentially infinite potential which is constrained by issues of human inertia, capital formation, social responsibility and economics. Most important is the issue of economic justification. No one needs a robot! Factory managers buy robots only if they expect economic benefit.

Right now economic justification usually requires two-shift operation. And it does make good sense to utilize any capital asset as intensively as possible to maximize the return on the investment. But the industrial robot is rapidly converging on the time when single shift operation will become justifiable. Economics of scale are being realised as robot production volume increases. So too, cost of control hardware is still coming down.

Even though software cost is escalating, it can be amortised over larger quantities. By and large robotic functionaries will inexorably keep gaining economically over their human counterparts. Normalising, to eliminate the impact of inflation, I would predict that throughout the next decade robots will log enhanced economic advantages over human workers at a compounded rate of 5 to 7% annually. Clear single-shift justification should occur in the 1986–1988 period.

There is yet another factor, and that is the place of the robot in service industries, as a hobby, as a pet, as a domestic. The justification then is no longer solely economic. The robot in household service may be on call 24 hours a day and yet enjoy just as much idle time as do human servants. The masters of such robots will be looking more for comfort than for throughput.

One last prediction: Well before 1993 IFS Publications will see the need for changing the name of its pioneer journal, *The Industrial Robot*. Robotics will no longer be bound solely to industry. Brock will not brook a limited horizon!

Better things for a man to do than be a robot

George C. Devol (below), who wrote this article, is the father of Unimation. In 1954 Devol patented 'universal automation' or 'unimation' which involved 40 patents and eventually became the basis of Unimation, the world's largest maker of robots

CENTRAL to my belief in the potential of robotics to cure a nation's industrial malaise is the tenet that in our industrialised world, everything ultimately depends on labour. Robots are the first direct replacement for human labour.

Indeed, the word *robot* is derived from the Czechoslovak *robota*, which means 'forced labour'. Thus, one could think of robots as a new kind of slave – absent the negative connotation; robotic slaves *should* be utilised for humane reasons.

As has been pointed out many times, robots, which can eliminate or minimise dangerous, dull, and repetitive industrial operations from the human workforce, will have a profound influence on the way industry – and society – is structured in the coming years.

For example, the Society of Manufacturing Engineers jointly with the University of Michigan forecast that by 1995 worker injuries in factories will be reduced by up to 41% as a 'result of robot installations'. And contrary to some doomsayers, the June 1982 report of the combined group says that the 'vast majority' of workers displaced by robots will be employed in new positions such as robot programmers and technicians. They forecast that only 6% of displaced workers will actually be terminated during this decade and only 2% thereafter.

What makes robot users so enthusiastic about robots? The *Robotics Survey* of the Carnegie-Mellon University (1981) ranks the reasons that make today's robot-using industrialists bullish. In rank order: improved product quality is foremost, followed by reduced labour cost, elimination of dangerous jobs, increased productivity, increased product flexibility, reduced materials waste, compliance with the regulations of the Occupational Safety and Health Administration, reduced labour turnover (surprisingly . . .), and reduced capital cost. These results pretty well say it all.

One of the problems with introducing robotics to a mature industrial complex is that there is a lot invested in existing, nonrobotic ways of doing things. To optimise robotisation, the manufactur-

ing system must suit robotic capabilities.

To a robot, a human-run assembly line is hopelessly 'mixed up'. Robots and people do not work the same way; machines have little if any discretionary judgment.

For example, consider how parts are recognised by people. We see in three dimensions and use subtle clues such as shadows and reflections to determine the shape and orientation of parts. But today's robots have very limited vision capabilities and three-dimensional vision is just now beginning to reach the stage of commercialisation. Thus, a batch of parts might have to be light-coloured on one side and dark-coloured on the other to permit primitive robot vision systems to determine whether a part is right side up or not.

In another example, picking parts from bins, humans have very little difficult in orientating their hands or tools and selecting particular parts to be moved. But bin-picking is an exceedingly difficult task for robots, and despite the interest shown by some highly motivated researchers, I would recommend that bin-picking be eliminated and parts fed to robots only as they are needed.

No doubt many improvements will be made in robotic sensing: sight, touch, and speech, for example, as well as in adaptive control (the optimisation of operations *as they are performed*, based on sensory data and predetermined goals) and schemes for linking individual robot 'work cells' together through computers. Such improvements will come gradually over the next few years.

And I have a suspicion that among the basic improvements to robotic machinery itself will be the attainment of much faster operating speeds than are currently possible. Flexible construction could be the means of accommodating the forces of rapid accelerations and decelerations that would be produced by such rapid movements. Sophisticated accelerometers would, through feedback loops, enable microprocessors to achieve high degrees of accuracy despite variations in the robot's actual structural dimensions during such movements.

In addition, robots will assuredly be given mobility, which would be of great use say, inside nuclear reactors, on the ocean floor, in spacecraft, and even in warfare. Imagine a robot that could simply be placed inside an infantry tank and operate it much as a human soldier!

But we should not be impatient to see such developments. I estimate that the stage of robotics technology today is roughly comparable with the state of computer technology some 30 years ago. Even today, different makes of computers cannot communicate with each other efficiently. With this in mind, we should not be surprised that robots from different manufacturers and nations are incompatible. We should not expect standardised designs to be developed for decades.

Meanwhile, it is important to be practical. We should take refuge in the fact that very crude systems can accomplish an awful lot. Elegant capabilities are nice, but often unnecessary. The Japanese have done marvellous things with simple pick-and-place machines because the manufacturing systems utilising such devices have been fine-tuned for their minimal capacities for flexibility. Ultimately, the only limit on capabilities is their cost.

'A stand-alone NC device'

The Robotics Institute of America set forth a formal definition of robot, which was recognised internationally at the 11th International Symposium on Industrial Robots in Tokyo, October 1981. That definition, while extremely formal, does differentiate between genuine robots and machines that appear to be robots, but which do not quite make the grade. 'A robot is a reprogrammable multi-functional manipulator designed to move material, parts, tools, or other specialised devices through variable programmed motions for the performance of a variety of tasks.'

My own definition of robot is perhaps more direct: A stand-alone, numerically or digitally controlled piece of equipment that can pick up and position almost anything in almost any direction and in almost any plane; it must be able to receive and use information from computers and to give information to computers.

'Algorithms only now being developed'

If robots are to work together, the mathematical complexities involved in their networking in a typical factory setting will call for highly sophisticated analysis (software) to accommodate the many possible variations and problems as well as powerful computers (hardware) to make it all happen.

The algorithms upon which such software is based are only now being developed, for example at the Massachusetts Institute of Technology's Laboratory for Information and Decision Systems. A recent report from that laboratory notes that problems such as calculating the average production rate, in-process inventory, machine utilisation, and per-part-costs for transfer lines and assembly networks that are subject to randomness (breakdowns, for example), limited between-stage 'buffers' (storage facilities for in-process parts), and to routing schemes of many kinds is 'extraordinarily complicated, and require major advances in the state of the art in systems engineering, operations research, and decision and control sciences' (S. B. Gershwin et al., 'Progress in Mathematical Methods for Manufacturing Systems', M.I.T., Jan. 1981).

I believe that there are better things for humankind to do than to be robots themselves. It is a horrible concept to conceive of a man keeping up with a machine. But if we plan wisely, we can utilise these machines to put our nation back on the productivity track and leave to posterity a world, which although highly industrialised, will be far more pleasant a place to live and work in than it has ever been.

Man's inhumanity to man: Now the robot takes man's place

The Industrial Robot has an exclusive interview with Johan Espeland, production manager at Gustafsberg Fabriker, one of the first firms to use a robot

GUSTAFSBERG Fabriker is owned by the Cooperative of Sweden and is the largest manufacturer of sanatory ware. It has an annual turnover of about £55 million. Gustafsberg Fabriker was one of the very first companies to use robots for industrial coating. Their first industrial robot was introduced in 1969 by Johan Espeland. This was the Trallfa robot. Here he looks back to those early days and also makes some comments on the future.

You bought one of the very first industrial robots from Trallfa in Norway back in 1969 when hardly any industrial robots existed. What were the reasons behind this?

JE: Well, actually I wasn't looking for a robot for coating but for handling our tubs. They had to be lifted from one conveyor to another and we had real difficulties to get people to do that job which required 8-10 hands and is boring and heavy.

'They weren't very interested in my problem'

And why Trallfa?

JE: In a technical magazine I learnt about Trallfa and I went to Bryne in Norway to see them. There they weren't that interested in my handling problem but more about using a Trallfa for primer coating application. And I decided to try, because we had the same problem in that area with high labour turnover of coaters and ever increasing difficulties in recruiting skilled coaters.

What is your experience with this first industrial coating robot?

JE: It was a wild one. It painted our operators as gladly as it painted our tubs. We really had problems with the control system to get the coat when we wanted it and where we wanted it.

What do you consider to be the most striking difference between this first Trallfa and your latest one?

JE: In the first place there is a price difference. We paid about four times as much for the latest one than we did for the first one. But with inflation and a wider range of functions the real difference might not be that big. The

Above: Manual spraying of baths. Below: Quicker by robot.

difference lies in the memory capabilities, the programming possibilities and the control functions.
When did you buy your second Trallfa?
JE: The next year which was 1970, we bought another two Trallfas for primer coating. In 1971 we bought another two for top coating. Today we have three Trallfas for primer coating, five for top coating, two for handling and an 11th one merely for the fun of it. This one is used for our own experimentation and also as a reserve.
What do you consider to be the most important technical improvements on the Trallfa during these 14 years.
JE: The memory function of the robots.

'Robots eliminate jobs and job opportunities'

When the diskette came, it became much easier for us to store programs. Besides we can now run two hours programming cycles instead of 115 seconds with the first SCT (Single Cassette Tape-reader) memory. On the very first robot the components couldn't withstand the heat in the plant, but by 1971 this problem had disappeared.
What about the attitudes towards industrial robots?
JE: There is no doubt about it that a robot eliminates jobs and job opportunities. However, no objections are raised when it comes to this type of heavy, boring jobs that might even be hazardous to health. The Unions have been very positive to the automatisation of such jobs. The reasons for a robot installation are about the same today as they were in 1969. The labour market, though, is harder today and the Unions are a bit reluctant to automatisation as such.
Besides getting rid of jobs where it is difficult to recruit people, have you gained some advantages in production?
JE: One of the advantages is that all tubs are identically coated. It is also obvious that a man just can't work at his best solidly throughout a working day. He will get tired after half an hour, then he needs a rest and another coater has to take his job. It is also obvious that by the end of the day neither one of them have the same spirit as in the morning. A robot installation leads to increased capacity of the coating operation. Still another factor which we haven't measured exactly, is that we also save enamel with the robots.

You asked earlier about attitudes and there is one peculiarity I would like to mention. Those who at first were most reluctant to accept the robots were the foremen. They could always shout at a slow worker to speed up or to do a better job, but it was more difficult for them to get a Trallfa to behave. Their attitudes towards robots were hardly positive in the very beginning.
Is it still so?
JE: Not at all. Those of the foremen and the coaters left in the plant are operators and now they have a technically far more interesting job than they had before.
Do you see any problems caused by the automatisation?
JE: Well, if you mean problems we didn't have before, we now have to have training programmes on how to coat manually in order to be able to program the robot by the teach-in principle.
Who gets the benefit of the automatisation?
JE: It is easy to figure out who is winning and who is losing in this on-going process of automation. The pioneering companies making the right investments leading to lower cost will have a higher profit. Those are the winners. The losers on the automatisation are the companies who are forced to do something about their costs due to price competition. They seem to be stuck with permanently higher costs.
What are the innovations you would like to see in robotics?
JE: In the first place I would like to see more special tools developed for Trallfa robots. Then I would very much like to see a supervision system developed. A system that gives us a good overview of the robots, the conveyor and the whole coating operation. One other very important factor is the possibility of being able to control the coating process. There is a tendency that the line will be more complex in the future because more aiding functions are added and must be integrated. This will effect the control, and measuring systems will be needed to control and optimise the operation.
How would you like to summarise the past years with your Trallfa robots?
JE: The Trallfa robots have resulted in a big step forward for our coating operation. For enamel as well as for paint I don't think that any technical innovation has meant so much for the development in this field as the Trallfa robots have done.

Guidelines for the profitable installation of a Trallfa robot in industrial painting operations.
In what kind of painting operations will an installation of a Trallfa robot pay-off most rapidly? Here are some guidelines based on hundreds of installations during the past years.

- ☐ If objects are transported by a conveyor it is more likely that a Trallfa will be profitable.
- ☐ The higher the number of painters working manually the more profitable a Trallfa will turn out to be.
- ☐ The higher costs required for reaching the ppm limits, the higher are the possibilities for a profitable Trallfa operation.
- ☐ The more difficult and the heavier the manual painting operation is, the more necessary Trallfa robot installation becomes.
- ☐ The more difficult it is to find skilled, manual painters, the more necessary a Trallfa installation becomes.
- ☐ The more frequent absence of the manual painter, the shorter the pay-off time for a Trallfa.
- ☐ The longer the production series, the more profitable a Trallfa will turn out to be.
- ☐ The more mixed the production series are, the more profitable a Trallfa will be. The number of objects can be as small as 10 units in each series.
- ☐ The more layers painted, the more profitable will a Trallfa be.
- ☐ The less the productional slack in the painting operation, the higher the possibilities for a successful robotisation.

British Robotic Systems Limited

Automated vision sensing for manufacturing industry

Our Products

British Robotic Systems Ltd (BRSL) provides a range of products and services for industry involving the use of intelligent sensors, such as vision, for the automation of a wide variety of manufacturing processes.

The Benefits

The financial return from the use of intelligent sensors can be very substantial indeed. BRSL is building a commanding international lead in the application engineering necessary to ensure that these returns are achieved in practice.

What are Intelligent Sensors?

Intelligent sensing means that robots, previously limited to a series of pre-arranged repetitive operations, can now react and adapt to changing circumstances. For example, a robot equipped with vision sensing can be given 'real time' signals to guide it to pick up a moving object - irrespective of the object's orientation. Indeed, using vision a robot can first recognise a component (say on a conveyor amongst a group of random objects), then check for quality and establish the orientation and finally, if satisfied, place it in readiness for the next operation in the manufacturing sequence. All this can be carried out under the normal working conditions found on the factory shop floor.

What we can do for you

BRSL's staff are primarily application engineers skilled in applying intelligent sensing to manufacturing control systems. The range of applications is enormous, extending from single activity tasks such as reliable 100% inspection to the linking of many complex, but often conventional, items of manufacturing plant under one integrated control. BRSL supply complete turnkey systems where, for example, the intelligent sensor acts as a controller for a number of components such as robots, machine tools, conveyors, quality control mechanisms and data loggers. BRSL is one of the few companies who have supplied operational vision sensors for a Flexible Manufacturing System.

How we do it

The AUTOVIEW* image processing system is at the heart of the control systems supplied by BRSL. AUTOVIEW is a generalised image processing computer package which is capable of examining real time camera images over a range of 256 light intensities. This is three times more sensitive than the human eye. AUTOVIEW has a range of over 200 commands available and is constantly being updated by new in-house developments and published work. Typically one industrial application of AUTOVIEW may use a set of only four or five commands to arrive at a decision making sequence, others may use twenty or more. The scope and versatility of the command combinations ensures that almost all applications can be quickly developed.

Sequence of Events

BRSL AUTOVIEW systems are implemented in three stages. The Development stage involves studying the various image processing options to select the most practical and cost effective sequence. This work is carried out either in BRSL's or the user's laboratory using an AUTOVIEW VIKING Development System. The next stage, Engineering, includes the integration of these results into the overall plant design, component procurement (robots, computer, conveyors, tooling etc) and works pre-commissioning of the assembled system. The final stage, Site Installation and Test, includes customer training, handover and ongoing support.

Typical Users and their Applications

BRSL's products are already in use in many industries ranging from automobiles and aerospace to pharmaceuticals, food processing and electronics. AUTOVIEW vision systems are used for a wide variety of tasks including on-line quality control, inspection and measurement, robot guidance and the many diverse activities needed for manufacturing automation. BRSL's AUTOVIEW systems are particularly successful for routine tasks, such as repetitive inspection, where continuous 100% human concentration is impractical. Also since it is so easy to teach AUTOVIEW sensors new tasks, BRSL'S products are well suited for small and medium batch production lines where the operations are frequently changing.

Who are we? - Some Background

The commercial progress made by BRSL during the period 1979-1981 was recognised by the British Technology Group who decided to provide additonal backing for the company in 1981. With BTG's support BRSL has built up its management and engineering teams and has adopted an ambitious research and development programme in order to keep the company, and its engineers, at the forefront of intelligent sensing technology.

How we work with our Customers

For the design and implementation of new turnkey projects, BRSL's engineers normally work with user's engineers. In this way we can effectively combine our joint skills. Alternatively, where the user's engineers are not readily available, BRSL engineers are able to supply fully integrated intelligent systems as "black boxes" where the user need have no concern about what takes place internally.

Summary of our Application Services

BRSL will design, install, commission and support the equipment on site and teach users how to operate it. In addition, BRSL can supply plant and production engineering and project management for full turnkey systems. This includes related engineering tasks such as the design and supply of jigs, fixtures and handling equipment. And, of course, BRSL engineers are available for rapid call out whenever maintenance is required.

For further information contact
British Robotic Systems Ltd
9-11 Grosvenor Gardens
London SW1W OBG
Tel: 828 4137 & 403 6666

The first robots
How they were justified

William Marshall, works manager at Doehler-Jarvis, recaps how the first robots were justified

HAVING utilised robots in our die casting operations at the Pottstown Plant since 1964, I have often been queried by potential robot users as to how the initial expenditure for a robot was justified. Not wishing to be cheeky, at least while engaged in a personal conversation, I usually replied by stating justification was based on anticipated increase in productivity. Certainly, this was the desired end.

The truth of the matter is something else, at least if by justification one is referring to the method of preparing a formal appropriation request.

My preferred answer would have been 'A strong conviction that robots could increase productivity.' I could not prove it on paper – at least not at first. As a matter of history, the initial request to purchase a robot (despite mathematics that alleged productivity gains could comfortably justify the cost) was denied not once but twice.

It was then that a loophole in corporate grants of authority became the means of acquiring our first robot. We found that renting equipment did not require approval by upper management so we simply rented the robots. By the time the loophole was closed, the robots were in operation and justification for the 40 robots now in the plant was well documented.

As an aside, regulations imposed by corporate structure do, at times, impede entrepreneurial type thinking by accepting only those ideas having no risks attached.

Getting back to the proven justification, the basic payback was increased productivity through greater yield of good pieces. With part orientation a by-product of the die casting process, teaming with a robot became an easy way to eliminate variables in the process caused by operators. The use of robots also clearly pointed out modifications of equipment and tooling that produced still greater number of good pieces with accompanying cost savings.

Today robots are not only accepted but are heralded as the panacea for declining productivity in the Western hemisphere. With such euphoric acceptance, robots are far easier to justify today. I also perceive that robot applications will not always be completely valid and that a used robot market will develop.

Our own operating experience with robots has been and continues to be wholesome. Robot capabilities and reliability have increased steadily and the relative cost of robots has steadily decreased. Thus problem areas, especially those of reliability, have diminished over the years.

Robots first acquired in 1964 have been replaced because of technical obsolescence. However, we still have 25% of our robots in the plant with 10 or more years of service. Normally, a robot year is 3,500 to 4,000 working hours.

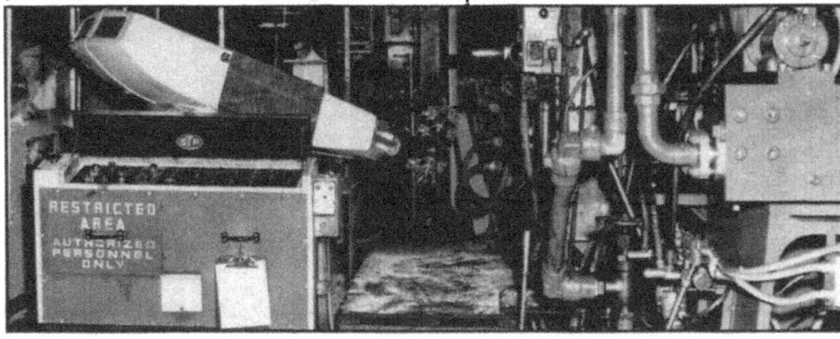

A Unimate at Doehler-Jarvis.

Maintenance, once thought reserved to white-coated specialists, now is accomplished by a regular plant maintenance force trained and updated in sophisticated electronic and hydraulic technology as well as diagnostic skills.

The acceptance by the work force, at least in our plant, has been not only co-operative but clearly positive. They see robots as a means of providing additional security to their own jobs. They have also witnessed for themselves that robots do relieve them of jobs having something less than an ideal work environment and have in turn given them an opportunity to get a piece of management responsibility by being bosses of the robots. The work force now has a true opportunity to increase productivity beyond its own muscle power.

Robots have been a most effective catalyst in reshaping die casting from an art to a process that can be placed under statistical control. Robots still will be used in a stand-alone manner, despite the fact that significant advances have been made to integrate the die casting process with automated peripheral handling and secondary operation equipment.

I say this for several reasons:

- ☐ Industry is decentralising into smaller, easier controlled units of manufacture.
- ☐ Computer Numerical Controlled machining and processing vastly reduces set-up time over previous methods. Lot quantities can now be smaller.
- ☐ Inventory costs have been targeted as a needless unproductive expense and forced a concerted drive to cut to levels previously considered not prudent.
- ☐ With inventory stocks low, incoming quality is an absolute necessity.

All of the above mandate that the die caster, like all basic component makers, must be able to handle small lot quantities with minimal lead time or suffer burdensome inventory costs of safety stocks. The die caster can better meet this challenge with flexible stand-alone robots than with the integrated peripheral equipment concept.

Holding that the future will unravel as prophesied above, I believe the most urgent need for development in robotics is that of sight: that is that a robot will have the ability to 'look' at pieces randomly jumbled in a tote box, pick them up and orient them for processing in a subsequent operation.

Until such a break-through occurs, use of robots in die casting will continue to expand, perhaps more slowly, in the secondary operations area by teaming robots with operators. Here the emphasis will be on using operators for judgmental portions of tasks (inspection, measuring, etc.) and have robots do the menial tasks.

· A Decade of Robotics ·

The Industrial Robot

Let the application choose he drive

Joseph Ray, chief engineer of Cincinnati Milacron's industrial robot division, looks at hydraulic versus electric drives

WHETHER POWERED by hydraulics, electric or pneumatics, drive systems used on industrial robots may be categorized as either direct or indirect drives. Each type of drive system has certain advantages and disadvantages, with the requirements of the intended applications determining which drive is appropriate.

For instance, in applications in which the robot must carry heavy payloads and withstand heavy impacts, a direct drive system, using hydraulically-powered actuators, can be most effective. But where the work involves lighter payloads, or where greater stiffness is required, an indirect electrically-powered drive system might be preferred.

Direct drives are characterized as having no mechanical linkages between the input power side and the output shaft or slide. Hydraulic or pneumatic cylinders and single or double vane rotary actuators are typical examples of direct drives. When these devices are combined with a power source such as high pressure hydraulic fluid, they become relatively small packages capable of exerting large forces and moving robot arms at high speeds.

Direct drives have other advantages: they are simple; they require little maintenance; they are durable; they are capable of easily being protected from shock or overload conditions; and they are backlash-free and of reasonable cost.

There are, however, disadvantages: Although hydraulically powered direct drives do provide high power from a small package, the package is still substantial, and since these packages are located at the robot joints, they become part of the payload that must be carried. The drives have a tendency to be compilant and although this may be advantageous in some material handling applications, the compliance leads to an undesirable amount of overshoot if short, fast moves are required. Finally, since most of the drives are hydraulically powered, they are susceptible to leakage if not properly maintained.

Indirect drives are characterized by having some mechanical linkage between the input and the output shaft. These linkages may take the form of ball screws, gear trains, harmonic drives, chains, belts, and so forth. In almost all cases their purpose is to multiply the force or torque of the driving device to a level capable of moving the robot arm and its payload.

Since most of the indirect drives have a high gear ratio (50 to 100:1), they normally exhibit high stiffness. This characteristic is useful in applications such as drilling and routing and applications requiring short, fast moves.

Indirect drives allow the use of dc servo motors as the power source. These motors normally do not have the torque necessary to drive the axes of a robot without amplification. The dc servo motor has many advantages, especially for smaller robots. The drives are quiet, very responsive, and considerably more efficient than hydraulic power sources.

However, primarily due to the necessity of drive trains to amplify their torque to a useful range, they tend to be more expensive. And, since they are highly geared to generate the necessary torque, they tend to operate at high speeds. This limits their use on high payload robots requiring high speed motions.

Repeatability is one of the most important specifications associated with industrial robots. Since the linkages within an indirect drive are subject to wear, provisions must be made to compensate for this wear or backlash in order to maintain the specified repeatability. Direct drives, by their nature, do not have backlash problems.

On many occasions the environment dictates the power source to be used. If the machine is to be intrinsically safe, as is necessary for spray painting, then hydraulics or pneumatics are the only alternatives. If the application can stand no contamination by lubricants, then electric is the solution.

Large heavy-payload, high-speed robots will continue to be direct drive hydraucially powered for some time to come, and most of the smaller, lower payload machines will most likely be electrically powered. No single type of drive system will be appropriate for all applications. The intended applications will continue to dictate which system is most suitable.

Spot welding car bodies by hydraulic robot.

The importance of configuration

Ole Molaug, who developed the Trallfa robot has come up with another design (below), this time for Graco

I TOOK my first step in robot development in 1964, when I suggested that a spray-painting robot should be made to paint wheel barrows for Trallfa, Nils Underhaug. At this time I was employed by Jæren Automasjonsselskap, (Jæren Automation Company) which was a consultancy firm owned by, and working for, ten smaller Jæren companies.

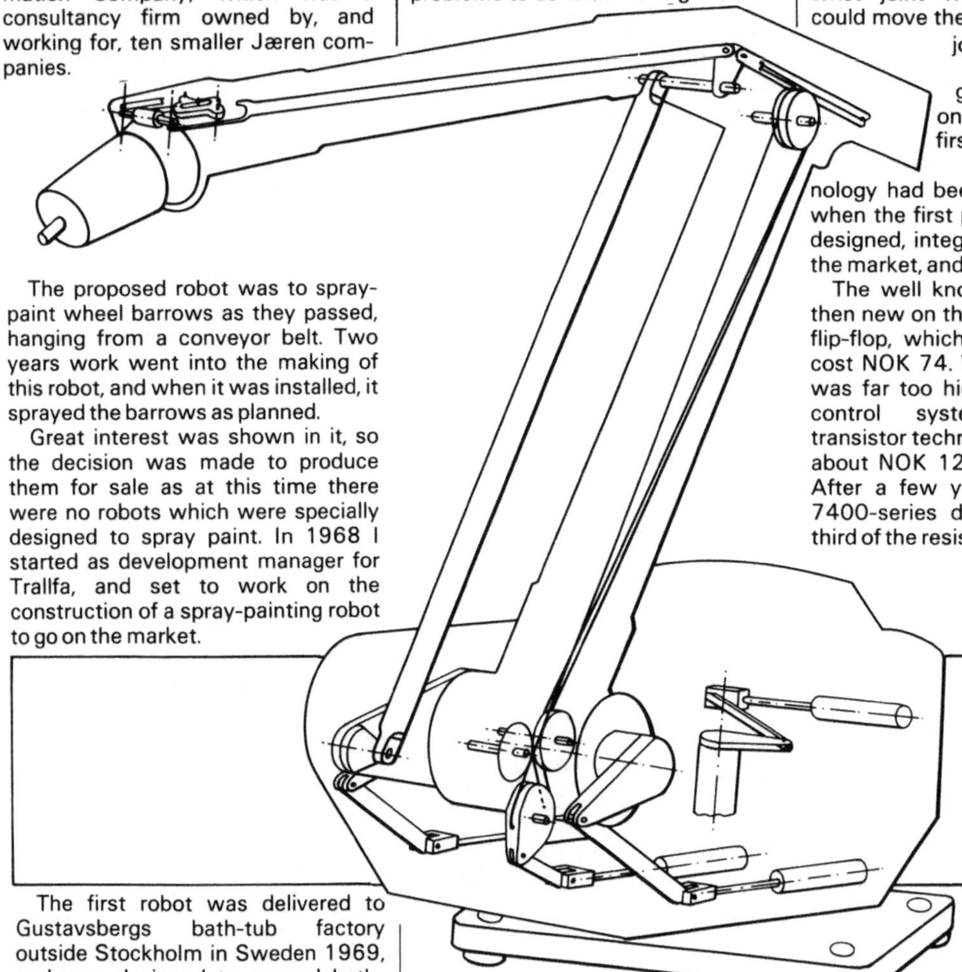

The proposed robot was to spray-paint wheel barrows as they passed, hanging from a conveyor belt. Two years work went into the making of this robot, and when it was installed, it sprayed the barrows as planned.

Great interest was shown in it, so the decision was made to produce them for sale as at this time there were no robots which were specially designed to spray paint. In 1968 I started as development manager for Trallfa, and set to work on the construction of a spray-painting robot to go on the market.

The first robot was delivered to Gustavsbergs bath-tub factory outside Stockholm in Sweden 1969, and was designed to enamel bath-tubs. This robot had five degrees of freedom.

The spray-gun could be moved within a three dimensional pattern, and from where the wrist-joint could be operated in two different directions. The spray fan on the spray-gun could not be turned.

This robot was by no means perfect, but it did a good enough job to warrant the investment in several others for this factory. This particular job that the robot carried out was in fact a very boring one, and it had been difficult to find people to stick it out for very long.

The robot was later used for spray-painting, but there were a few problems to do with making the robot explosion-proof and protecting it against the high tension from electrostatic spray-guns.

We solved these problems by making the robot intrinsically safe by separating the hydraulic pump and manipulator, and after a while our own zenerbarriers were approved. The robot was then reconstructed so that the pump, which had formerly been located in the robot base, could be taken out, and the manipulator part could then be made intrinsically safe.

After a while a sixth rotary direction for the spray-gun was desirable. To facilitate this, a so-called flexi-arm, or wrist joint was constructed. This could move the spray-gun like a wrist joint, in three directions.

At this time great steps were made on robot control. In the first robot electrontubes and transistor technology had been used, but by 1968 when the first production model was designed, integrated circuits were on the market, and prices began to fall.

The well known 7400-series was then new on the market and a double flip-flop, which today costs NOK 2, cost NOK 74. We felt that this price was far too high, so we designed a control system with resistor-transistor technology, which cost only about NOK 12, for a double flipflop. After a few years the price of the 7400-series dropped to about one third of the resistor-transistor circuits.

When the robot was made intrinsically safe, the complete control system was reconstructed to the 7400-series. In the meantime the development of more comprehensive types of integrated circuits continued, and we used an Intel 8080-processor to store data on flexible disks and re-feed data to the robot. The servo system was still the same as the one constructed in 1968.

In 1978, progress led to the

construction of a completely new control unit, which used the very latest 16-bits microprocessor which had just come on the market. This work was continued after I left Trallfa in 1979.

I started my own development company in 1980, and one of the first projects I took on was to make a new spray-paint robot for Graco Robotics Inc. The control system in this robot is exclusively microprocessor technology and is set up to control eight servos. Six are for the manipulator itself and the two extra servos can be used if the robot is put on a slide base.

The need for this extra mobility arises when spraying the interior of car bodies, or stationary large objects which the robot cannot completely cover length-wise or height-wise. All in all, the control unit is constructed on the principle that it can be built on to, as far as the software is concerned, such as object detection, kartesic converting etc.

The construction of the actual manipulator was based on our earlier experience. First of all we wanted to

Fig. 2. Ole Molaug.

make robot programming with the teach-in method as easy as possible. This means a light-weight, low inertia and low friction arm.

One solution was to incorporate an additional teach-in arm as adopted by other firms. However, we did not want an extra training arm. Our goal was to make a truly light-weight low friction robot arm.

By using a computer at the problem analysis stage, we managed to devise a counter-balance system which functions almost perfectly in every part of the working area. This means in other words, that this robot arm can in fact be in any position, and still remain completely counter-balanced.

To reduce the bulk and inertia of the arm to a minimum, all servo cylinders and counter-balance springs were placed in the body. One of the stipulations was, that the robot should not take up more space in the spray-cabin than competitors' robots, but should still cover as big a working area.

By lowering the pivot point of the vertical arm, and by changing the pivot point of the whole robot arm, we achieved a 60% greater working area than the earlier robots had covered. The idea behind changing the pivot point for the whole robot arm was to achieve a flatter circular arc and thereby allowing the robot to better adapt itself to a passing conveyor-line.

Trying to look into the future, I would say that the spray-paint robots with 'teach – playback' systems will dominate for many years to come, at least for spraying complicated parts. For simpler objects automatic programming will be possible involving for example a camera which registers the object in three dimensions.

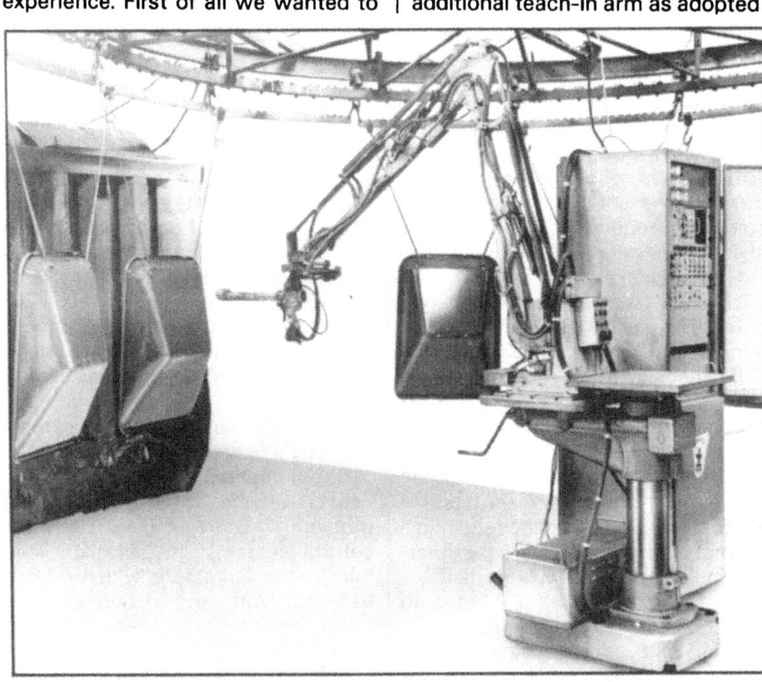

Fig. 1. Developing the Trallfa robot.

Fig. 3. The Graco robot.

Robot motion control: The key to success

Gordon VanderBrug is product line manager for Cybervision assembly products and system software at Automatix Inc while Ofer Gneezy is responsible for robot motion control in the same firm. Here they look at the technology

ROBOT motion control is based on control theory, kinematics, dynamics and digital signal processing. Although these areas are well developed, their application to robot control is not straightforward and the real time computation required to solve the equations of motion can exceed the power of a present day mini or micro computer. The equations can be simplified resulting in manageable computation complexity at the cost of performance deterioration.

In practice, the performance requirements are determined by the class of applications the robot is designed for, and that performance governs the acceptable degree of simplification. Trends in robot control can be discussed in the context of three representative classes of applications with progressively more stringent performance requirements.

The first class, Class 1, includes applications such as spot welding that require position repeatability but are insensitive to the trajectory the robot takes when moving from point to point. Robots of this class operate in joint space.

'Motion control is carried out in the servo loop'

To move the robot each degree of freedom can be commanded to go to its goal point at maximum speed. This results in un-coordinated joint motion. The robot traverses a complex trajectory that depends on the geometry of the mechanism, and the axes arrive at their goal points at different times. Axes synchronization can be achieved by scaling the velocity of each axis relative to the slowest axis.

To shorten the overall cycle time the robot can be commanded to move through some points without stopping. This is achieved by planning the transition between successive motions rather than decelerating the robot to a stop at the end of each motion. The result is continuous path joint motion.

The actual motion control is carried out in the servo loop. If it is assumed that all the axes are independent of each other, their inertias are position independent, and gravity loads and friction are negligible, then each axis is a simple second order system, and can be position servoed with position and velocity feedback.

The velocity feedback can be obtained from a tachometer and fed directly into the servo amplifier while the position feedback can be obtained from an optical encoder by a digital proportional servo loop. Some compensation (such as lag leed) can be added to the servo amplifier to attenuate the impact of the structural frequencies.

Thus, for this class of applications, sufficient performance is achieved at a very low computational complexity.

The second class, Class II, includes applications, such as arc welding, that require control of the trajectory along a predetermined trajectory. The trajectory can be a straight line, a circle or another curve.

Straight line motion is achieved by interpolation in Cartesian space rather than joint space. The interpolated points are then converted from world space to joint space. This conversion is the inverse kinematic problem.

The complexity of the equations varies with robot geometry, but for all but Cartesian robots is very high. For example, the solution for an articulated robot requires 13 transcendental function calls, 27 multiplications and 20 additions.[1],[2]

To achieve a good approximation of a straight line the solution has to be found at a very high rate (10 to 100 HZ), especially if the robot is moving fast. It is possible to precompute the solutions but this requires large memory space and prevents the alteration of the trajectory in response to sensory data. The representation of points can be in joint space requiring a kinematic solution to start the interpolation, or it can be in world space.

There are several advantages to world space representation: the location data is robot and tooling independent; the location data can be generated off line; and the task can be easily moved to a new area in the robot work envelope.

There are several ways to store world space locations: homogeneous transformations (4×4 matrices containing a location vector and three unit vectors that define the orientation), three position values and Euler angles defining the orientation, three position values one unit vector and a rotation angle about the unit vector, and others.

Finally, the third class, Class III, includes applications, such as small parts assembly, that require high velocity and acceleration motion and exertion of controlled force for part mating.

When exerting forces, a force sensor has to be added. It can sense torques in joint space or forces and moments in tool space.

The position servo can be modified to a torque servo. Because of the high performance requirements the non-linearities and cross coupling of the dynamics equations cannot be ignored. The resulting control law used in the servo loop is much more complex.

'One aspect of total robot control'

For high performance over the full range of motion the servo gains must change as a function of the position. For stability, the servo rate for most robots is in the range of 100 to 1000 HZ. Control of robots in this class poses severe computational difficulties even when a separate micro computer is dedicated to the servoing of each axis.

Besides these three classes of robot motion there are other trends.

Robot motion control is only one aspect of the total robot control problem. State-of-the-art robot control systems include high level interpretive languages which make it easy to write and debug complex robot programs.[3] These systems support the standard operators and control structures of a programming language, and in addition have structures to facilitate writing robot programs, such as data types for paths, reference frames and tool definition.

Modern robot systems also support interfaces to sensors, such as tactile, force and vision, and even provide the support of a programming language to write the sensory processing algorithms.[4] Interfaces from robot data bases to CAD data bases which result in substantially reduced programming time have begun to appear.[5]

In addition to the above trends, languages will include models of the part being processed and its environment. The interface between control and sensors will appear at multiple levels and the two functions will become more tightly coupled. Interfaces between robot and CAD data bases will be enhanced, and will be extended to other manufacturing data bases, such as process planning.

REFERENCES

1. R. P. Paul, Robot Manipulators-Mathematics, Programming and Control, MIT Press, Cambridge Mass., 1981.
2. C. S. G. Lee, Robot Arm Kinematics Dynamics, and Control, IEEE Computer, Volume 15 Number 12, December 1982.
3. J. W. Franklin, and G. J. VanderBrug, Programming Vision and Robotic Systems with RAIL, Society of Manufacturing Engineers, Robots VI Conference, March, 1982.
4. A. Reinhold and G. J. VanderBrug, Robot Vision for Industry: The Autovision System, *Robotics Age,* Fall 1980, pp. 22–28.
5. P. Villers, The CAD/CAM Link, *Industrial Engineering,* Volume 14 Number 4, April 1982.

It's the language that decides how well the robot performs

Peter Davey, robotics co-ordinator at the Science and Engineering Research Council, examines robot languages

THE sole purpose of a robot language is to make it easier to define the task a robot has to perform. Language is only one part of the complete software system needed to programme a robot off-line – that is, without requiring the actual robot and workpieces.

Off-line programming may never be required for the simpler robot applications; but will become increasingly necessary in future where an application involves:

○ Complexity, especially in assembly tasks.

○ Small batch production, where a major problem may be the expense and delay involved in making available during programming not just a complete robot cell, but also prototypes of each part it has to handle.

○ Links to existing CAD databases.

○ Danger to the operator during teach-in programming.

Robot languages are generally classified according to increasing power – or level of abstraction:

□ Joint level (individual settings of each joints must be specified).

□ Manipulator level (only the desired position and angles of the end effector need be specified).

□ Object level (only the movements required of the workpieces need be specified).

□ Objective level, in which the task need only be specified in the most general form, for example 'assemble carburettor type 4B'.

These levels of abstraction roughly parallel those in ordinary programming language which range from machine code, through higher languages such as FORTRAN, to planning levels such as PROLOG. No language much higher than manipulator level is so far on the market and even these have not been used very widely yet.

Commercially available languages include SIGLA (Olivetti 1978), HELP (DEA Pragma 1979), AML (IBM 1982). But the most widespread is certainly VAL, originated by Unimation around 1975.

VAL is an interactive language. It provides first a set of commands to the robot's operating system; second, a number of utility packages comprising an editor and packages for co-ordinate frame transformations, calibration, communication, and so on, and third, a set of instructions for defining motion of the end effector and control of sequences. These latter, found in all ordinary computer languages, provide the possibility of branching, looping, and so on.

Languages in the research stage are all aiming at *object* (or higher) level. They fall into two classes: first, those where a frame of reference must be explicitly defined in terms of position co-ordinates, such as AL (Stanford, 1977) and LM (Grenoble, 1980); and second those where frames can be defined implicitly relative to features of objects, such as AUTOPASS (IBM, 1977) and RAPT (Edinburgh, 1978).

AUTOPASS was the first attempting to reach *objective* level: it incorporated automatic planning of assembly in terms of movement of individual objects, each described by a geometric modeller. It did not enter production use partly because of the prohibitive amount of computer time it required.

The key ideas of RAPT are novel means of representing relationships (touching, sliding, turning) between individually modelled 3D objects. Currently, research is under way to incorporate these ideas in a complete system for off-line programming.

So, what functions will a future off-line programming system contain? Some consensus is beginning to appear about this. For example, Fig. 1 – based on a recent proposal for joint research in France, Germany and the UK – shows some encouraging resemblances to a recent proposal by a Japanese Interest Group of CAM-I. At the centre of Fig. 1 is the object level programming system.

Robot programming essentially concerns manipulating and observing a large number of 3D objects in space, so it is no surprise that the object level system must be served by a powerful body modeller. This in turn may draw on the factory's main CAD database (above) for some models of parts to be handled. Feeding into the object programming system will be a process planner.

This function is at the moment entirely undertaken by the manufacturing engineer, but later on it is expected that planning software will allow the man to specify what he requires in progressively broader terms. Intermediate steps, both in normal production and in the process of recovering from errors, will be undertaken automatically by the planner.

The man operating an off-line programming system must be able to visualise accurately just what will happen when his program is being executed. The right-hand block labelled simulation and visualisation is the software needed to do this.

However, there are still massive problems in making such visualisation on a VDU anything like as accurate as hands-on contact with the real robot. Display hardware to do this is one of two important areas needing work in the Man/Machine Interface (dotted line near the top). The other is development of appropriate command languages which will interpret the wishes of engineers well versed in different processes in terms which the automatic software can understand.

An important interface half way down Fig. 1 is that between software run off-line and on-line. The on-line software consists of applications programs generated by the modules already described, to run under the real time operating system of the robot itself.

These application programs will not only comprise the detailed sequence of trajectories required of the manipulator, but must also interpret data returned from sensors, and use that data to take corrective action.

Fig. 1 distinguishes the software modules by logical function alone. There is also an important physical distinction –how will the different software modules be mapped on to the physical hardware, for example, mainframe, supervisory computer, local area network, and robot controller. On-line software will probably be divided between a supervisory computer and a number of controllers

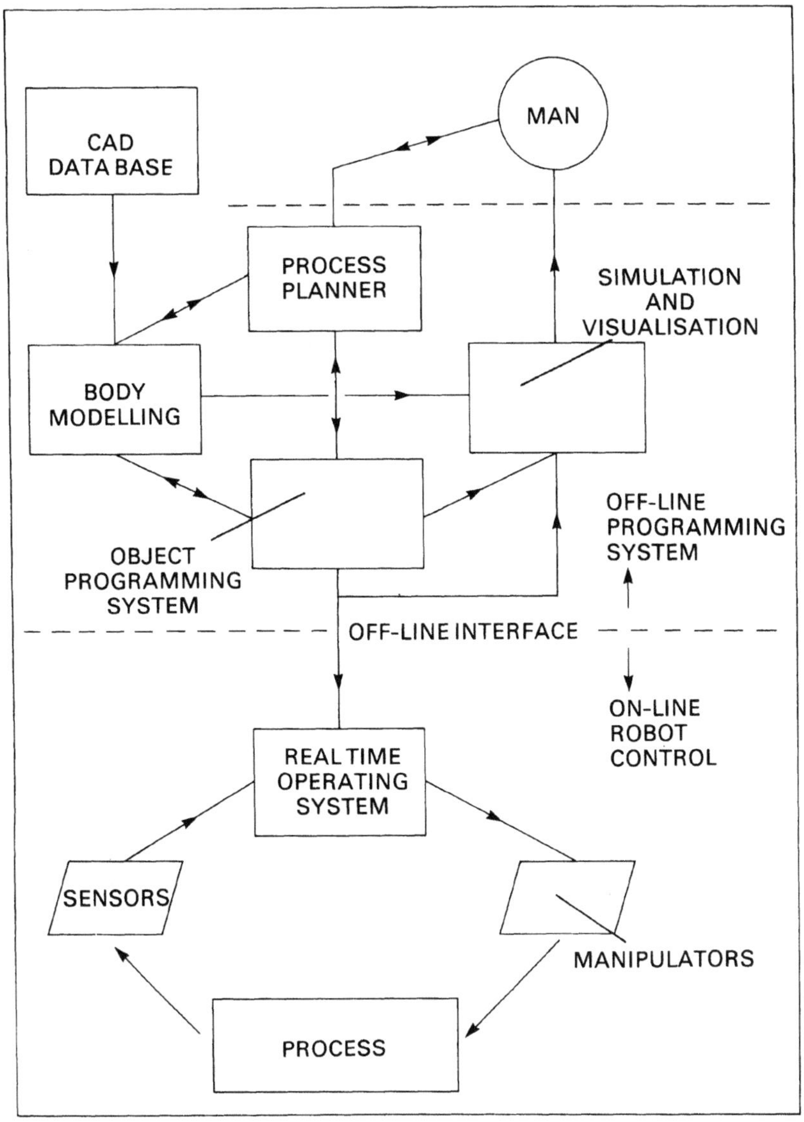

Fig. 1. Proposal for a robot language system.

for robots and sensors linked by a LAN.

There is now considerable activity in robot language all over the world. To production engineers trying to make today's robots work in present applications, all this may sometimes seem like a Tower of Babel.

Although much current activity is headed towards the eventual adoption of standard languages, it should be realised that complete off-line programming systems are still very much in the research stage, and it is too early yet for standardisation of languages by the usual standards bodies. The process will probably start with standards for nomenclature and later for important interfaces.

In the USA work has been under way for many years in the Language Standards Committee of ICAM – the US Air Force's Integrated Computer-Aided Manufacture Programme, mainly concerned with aerospace production. There is also work in the larger robot manufacturers, for example Unimation, Cincinnati and IBM.

Even more important may be that being triggered by some big users like General Motors which are getting increasingly worried about the problems of programming large populations of robots off-line, especially when these come from many different suppliers. MIT and especially Stanford, have been for many years the world leaders in the more academic aspects of off-line programming.

In Europe an early lead was given by two leading Italian robot manufacturers in providing manipulator level languages. In Britain University research work sponsored by the Science and Engineering Research Council has for the last two years been steered by SERC's Language Working Party – a group consisting mainly of representatives from industry and who help guide academic work in return for assessing early results on real applications. Comparable research is taking place in France and Germany also.

'Research crucial to better systems'

In 1981 a three-nation initiative including France, Germany and the UK, was set up by the relevant Government Ministries and has resulted in a proposal to carry out a one-year joint study in the three countries. This study, whose start is expected to be announced by March, will plan further research leading towards a standard language over the next four years.

In addition, through the EEC, there are two schemes which are likely to become involved in robot languages and their standardisation – the Computer Integrated Manufacture (CIM) project of ESPRIT, and the EEC Standards Working Group.

The Japanese are late entrants in the language race, and so far this has obviously not held up their application of robots! However, their track record in judging exactly when an emerging technology is becoming commercially important is perhaps the best of all. This makes particularly significant the recent formation of a Japanese Interest Group within CAM-I for Robot Language, which may well lead to further international development work.

I believe five R&D areas will be crucial in producing better robot programming systems over the next four or five years. These are:

□ Languages to create and maintain a computer representation of the robot's environment, or 'world model' while executing its task. To do this we need further development in ways to describe 3D objects (geometrical modellers) as well as ways to make inferences about relationships between those objects (geometrical reasoning).

□ Further integration of robot programming systems with the factory CAD/CAM database.

□ Improved processing of sensory signals (much of it making use of prior knowledge about the parts being observed – so called 'model driven' pattern recognition) as well as language developments to make it easier to define the correct response by the robot to sensors providing continuous information about, for example, position and orientation of workpieces.

□ Improvements in the man/machine interface. These will include better command languages 'tailored' to the needs of different processes as well as better devices for graphics input and output. In particular we need better 3D displays to present moving simulations of the robot, its workpieces and the appearance of signals from sensors –all in considerable detail.

□ Standardisation of crucial interfaces like that between off-line and on-line software. These will hopefully lead on to standardisation at least of primitives for higher level languages.

It seems that the majority of future research in robotics will be connected with software, and also that software modules are likely to be more universal than the actual robots and sensors which they control. While manipulators and sensors will tend to be tailored for particular requirements, major sections of the software, and especially software for off-line programming, will be independent of particular manufacturers and particular devices.

It follows from this that some of the most important opportunities for standardisation to ease the manufacturing engineer's task will eventually appear in robot programming systems. □

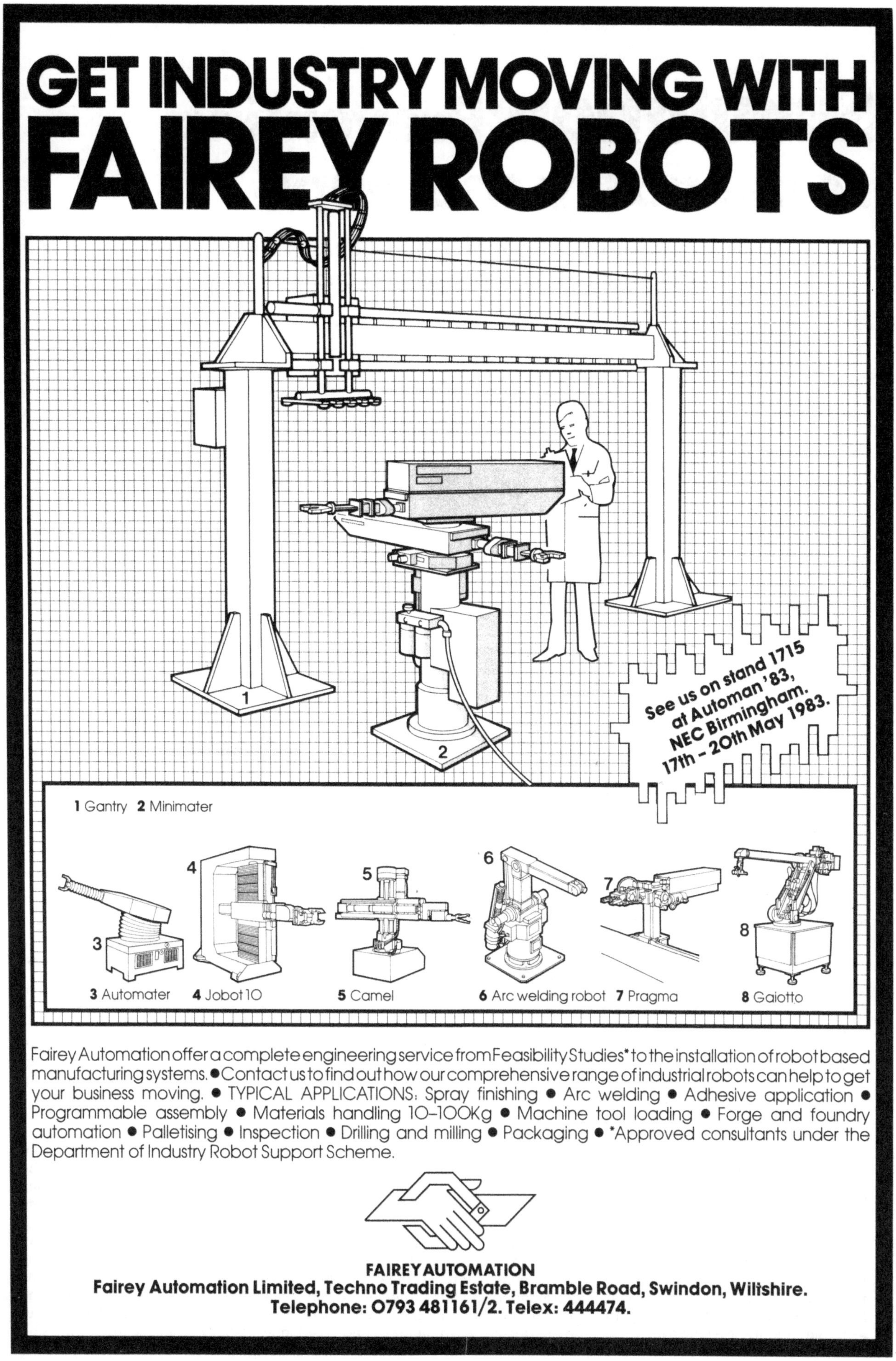
GET INDUSTRY MOVING WITH
FAIREY ROBOTS
See us on stand 1715 at Automan '83, NEC Birmingham. 17th - 20th May 1983.
1 Gantry 2 Minimater
3 Automater 4 Jobot 10 5 Camel 6 Arc welding robot 7 Pragma 8 Gaiotto
Fairey Automation offer a complete engineering service from Feasibility Studies* to the installation of robot based manufacturing systems. • Contact us to find out how our comprehensive range of industrial robots can help to get your business moving. • TYPICAL APPLICATIONS: Spray finishing • Arc welding • Adhesive application • Programmable assembly • Materials handling 10–100Kg • Machine tool loading • Forge and foundry automation • Palletising • Inspection • Drilling and milling • Packaging • *Approved consultants under the Department of Industry Robot Support Scheme.
FAIREY AUTOMATION
Fairey Automation Limited, Techno Trading Estate, Bramble Road, Swindon, Wiltshire.
Telephone: 0793 481161/2. Telex: 444474.

Simulation: Preventing some nasty snarl-ups

Dr Yoon Yong and Maurice Bonner of Nottingham University, UK, discuss GRASP, the robot simulation system

THE rapid growth of robot technology and the increasing complexity of applications means that engineers face new opportunities and problems. In particular the great flexibility of robot movement makes the problems of workplace and work task visualisation and planning more difficult. These problems are particularly acute when the robot is an element in a more extensive system, such as a flexible manufacturing system or a multi-robot installation which is linked with other machines and conveyors.

One solution to these problems can lie in the GRASP industrial robot simulation package which may be used to assist the design of robot installations. GRASP (Graphical Robot Application Simulation Package) is a computer aided design package designed specifically to plan and evaluate robot workstations. In addition to workstation layout, task definition and checking, the system also provides opportunities for evaluating robot safety problems and holds potential for integrating the CAD/CAM process. These facilities including the possibility for programming robots off-line are now discussed.

'To aid visualisation wooden models are used'

Both the dynamic nature of robots and the increasing complexities of modern installations provide CAD with ideal opportunities for involvement. Apart from the robot, for an installation, the integration of other equipment such as machine tools, fixtures and pallets has also to be achieved. Interaction between components of a system is normally difficult to evaluate manually using drawings in a two dimensional static mode.

To aid visualisation, wooden models are sometimes built. However this approach has limitations particularly if the elements are moving; for example a robot and a conveyor.

Interactive compuer graphics, especially if a three dimensional geometric modeller is available, can overcome many of these problems. Other advantages associated with the use of CAD include the ability to consider alternative options in design, allowing a better solution to be found. It also has the advantage of speed and flexibility for design changes. CAD does not by itself provide solutions and does not remove the onus from the designer but employed sensibly, it can provide improvements.

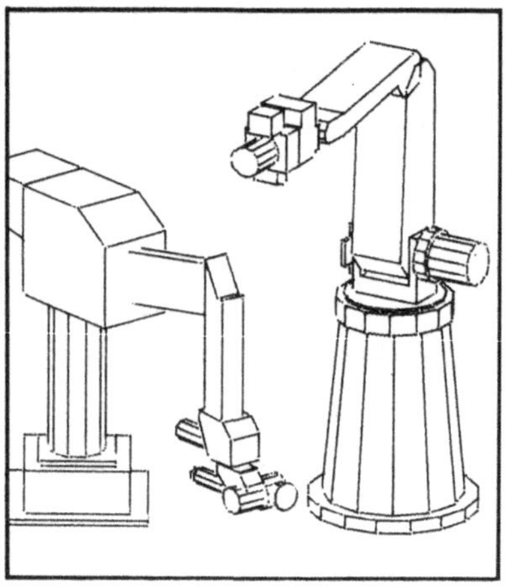

Fig. 1. Robot models of the Cincinnati Milacron T^3 and Dainichi Sykes BA 1440.

GRASP models are built up from a combination of primitive shapes which can take the form of cuboids, polyprisms, cylinders, etc. The user has control over the detail of the models built bearing in mind that the greater the detail the greater the computation required. Fig. 1 shows robot models of the Cincinnati Milacron T^3 and the Dainichi Sykes BA 1440 with hidden lines removed.

A comprehensive viewing facility within the system allows the user to drag and reposition models within the workplace, evaluating interactions until the user is satisfied with the layout. The kinematics of the system enable the simulation of robot motions to be carried out. Co-ordinate transformations provide solutions to the inverse control problem. Hence the location and orientation of the end effector of the robot can be specified in cartesian co-ordinates and the computer determines the intermediate linkage position.

Programming the robot model in GRASP is a relatively straightforward task and the user can do it in two ways. The first is analogous to the teach method used in programming industrial robots. The end effector of the robot model is dragged to the required position either with a light pen or via the keyboard, and the orientation of the end effector is then entered. In the second method, an entity representing an axis system is created and placed in the desired position and orientation in the workplace. The end effector of the robot model is then 'homed in' using the GRASP 'ONTO' facility.

A sequence of moves can be built up either by using several axis entities positioned around the workplace or moving a specific one around. This method of programming is advantageous particularly if input data to GRASP is supplied from a digitised format. It resembles that used for industrial robots when program moves are entered through the control system rather than being taught and recorded. Fig 2a and 2b illustrate this second method.

'Interference checking is very useful'

The value of the simulated robot programs to layout design is enhanced by the 'TRACK' facility developed in GRASP. Sequences are stored with reference to an object entity (for example a workstation) within the workplace. This allows a user to re-locate the robot without the need to re-program the sequences. It not only saves time but enables other siting positions of a robot to be evaluated. Similarly other robots may be evaluated using the same task sequence.

Interference checking between objects can be carried out within the workplace. Presently, only static checking is available within the system. During the planning, the user may visually detect potential collision points. If this occurs, the collision

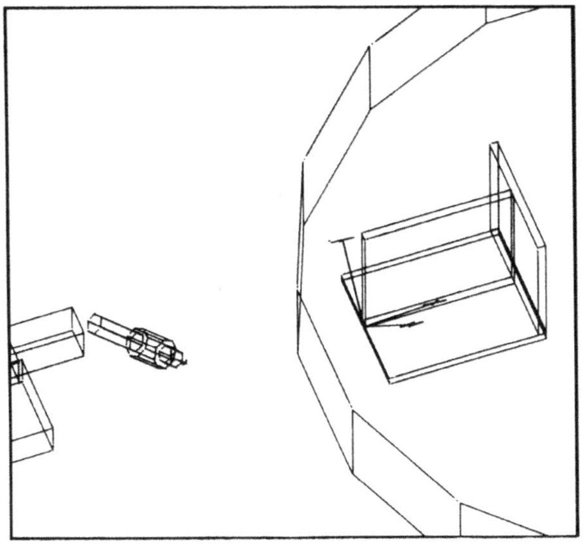

Fig. 2a. Programming with axis entity.

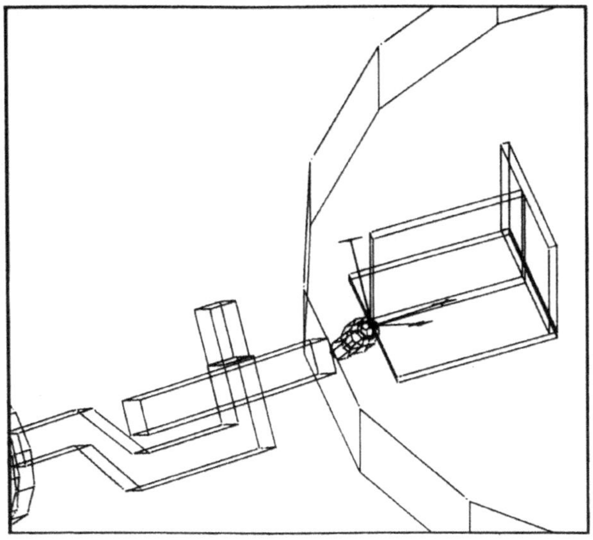

Fig. 2b. Robot tool homing onto axis entity.

detector should be turned on at the suspect points and any collision is hence reported.

Dynamic interference checking is currently under development. Once it is implemented, it will eliminate the need for preliminary visual checks and allow clash detection to automatically take place during the execution of the task.

Planning multi-robot installations from two dimensional drawings can be an extremely difficult task. For instance, actions may be required to occur simultaneously or complex interactions between moving objects may need to be evaluated. The GRASP 'EVENT PROCESSOR' overcomes these difficulties by relating events to an independent time base. An event could be an acceleration of a joint, a conveyor movement, a grip or a weld.

A collection of saved events with their associated parameters (for example starting times, durations, accelerations, etc.,) is formed into a process and by linking these together, complete tasks can be specified. In this manner, GRASP may be used to evaluate complex interactions between robots and other equipment within the workplace. A filmed computer animation of the task can also be produced.

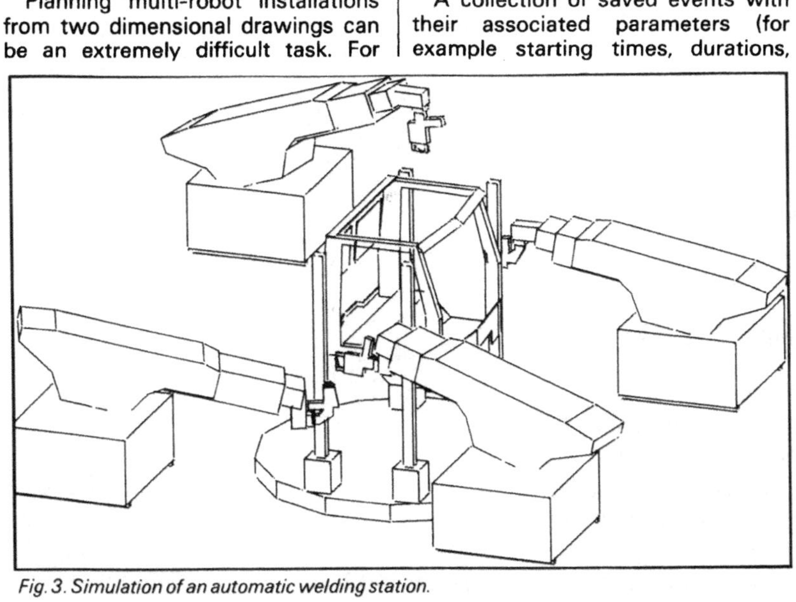

Fig. 3. Simulation of an automatic welding station.

'Structuring robot models is advantageous'

In GRASP a point on the last joint structure of the robot model is selected as the tool attachment point (TAP). To the TAP can be mounted the tools which the robot carries including weld gun, gripper, spray gun, and so on. Each tool has associated with it one or more tool centre points (TCP's) and the current TCP is then used in all areas of the system as the reference point for controlling the robot model.

The advantages of structuring the robot model in such a manner are, firstly each robot model can have more than one tool and TCP associated with it. These can be selected as required without the necessity of having to re-define or re-build the robot. Secondly the ability to vary the offsets of the TCP from the TAP

means that the system can be used in optimising the design of the linkage structure of the tools. This can be very useful if the robot has to work in areas where there are constraints on movement or limitations of access.

Putting a robot through its paces using simulation assesses the suitability of the robot for the task in mind. It can prevent an 'overbuy situation', that is, prevent the unnecessary use of an expensive and sophisticated robot for a relatively easy task.

On the other hand it can also determine whether the robot is capable of performing the task. For example, the effective working area of the robot is not its reach envelope. A task may require awkward orientations which may exceed the joint constraints of a particular robot. GRASP can check the reach and articulation capabilities of robots in performing tasks.

Linked to the system will be a robot library which will contain a large number of commercially available robots. A pre-selection algorithm may be used to initially select robots on the basis of the major attributes such as reach, load capacity, accuracy and cost. GRASP can then be used to assess a robots 'suitability'.

GRASP has been used for evaluating a wide range of tasks which have included spot and seam welding operations, brazing operations, paint spraying and materials handling. It is proving to be a flexible and economic method for evaluating robot workplaces. Fig. 3 shows the simulation of an automatic welding station.

Planning a robot installation by CAD also allows safety to be incorporated into the design process. The six axes of movement of the robot provide it both with flexibility and a large working envelope. Whilst this is clearly an advantage, it also makes the working envelope into a potential risk zone.

The guarding requirements around a robot installation is therefore one of the essential safety features which require careful consideration. Guarding is necessary to prevent entry by unauthorised personnel and also to prevent any accidental escape of projectiles during a gripper malfunction. CAD can help in designing the height and layout of perimeter guards.

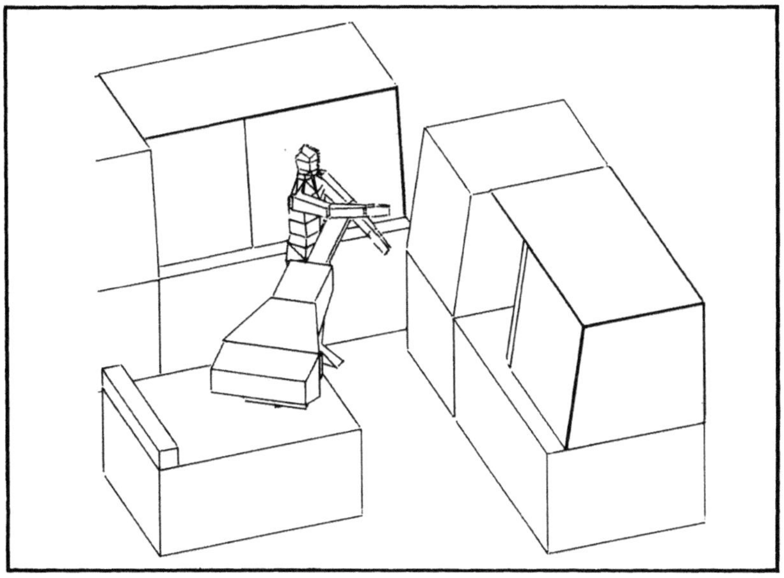

Fig. 4. An example of trapping in a robot cell.

The robot is sometimes capable of aberrant behaviour and this is particularly dangerous if it occurs in the presence of people. As robot installations normally have machines closely sited to each other, there are bound to be areas which can become potential trapping points. Where possible these trapping points should be eliminated by design.

In GRASP, the SAMMIE man model[1] can be used to simulate posture positions required for maintenance and programming. Fig. 4 shows an example of the man model trapped within a simple robot cell. Generally designing by CAD allows for a more systematic approach towards safety. A more detailed discussion of the subject can be found in Yong et al.[2]

The present method of programming industrial robots by the teach mode is clearly unsatisfactory from the safety point of view. For many types of robot applications (for example, arc welding), the operator has to work extremely closely during programming to obtain a satisfactory result.

Advances in robot technology have meant that there are now control systems which are capable of taking off-line programs. The implications of such developments on safety are that the essential man/robot contact time for programming will be reduced to that required for fine tuning.

Initial experiments for the transfer of off-line data from GRASP have been successfully conducted. Although there still exist many problems, we believe that this capability will be available to the industrial environment in the not-too-distant future. Off-line programming not only provides the industrial robot with additional flexibility but is the necessary link for integrating the CAD/CAM process.

The authors wish to acknowledge the support provided by the Science and Engineering Research Council in funding the GRASP project. They also wish to acknowledge the assistance given by their industrial partner the Production Engineering Research Association and Compeda Limited, co-ordinators of the project.

REFERENCES

1. M. C. Bonney, et al., 1980, 'Using SAMMIE Computer Aided Design System for Workplace Design', Institute of Management Services Summer School Cambridge.
2. Y. F. Yong, N. K. Taylor, M. C. Bonney, 'CAD – An Aid to Robot Safety', Robot Safety Seminar, November 1982, University of Nottingham.

Spot welding: The classic case for the quality robot

Stefan Müller, head of KUKA's robot division, outlines the case for robots in spot welding

THE prospective users of novel production methods have always had a strong influence on the development and application of these techniques. This is also true where industrial robots are concerned. So it is not surprising that numerous fruitful impulses from a highly progress-conscious sector – the automobile industry – are to a substantial extent responsible for the performance and range of application of these production assistants having attained their present, most encouraging level.

Whilst the development of industrial robots is now proceeding at a steadier pace, their application is constantly on the increase. More than half of the industrial robots in operation throughout the world are to be found in the automobile industry, whose investment programmes guarantee that this proportion will be maintained or increased through to the mid-eighties at least. A great many of these robots are spot welding units used for welding car bodies.

'It is important to maintain user connections'

Anyone wanting to start up in this sector nowadays must be able to offer robot systems which match up to the tough demands of continuous operation on the production line from the very first day onwards.

If a robot manufacturer can produce systems capable of doing this, it will generally speaking have fulfilled two basic prerequisites for the successful manufacture of high-grade industrial products: it will have substantial development potential at its disposal and it will promote productive communication with the users.

To derive maximum benefit from this cooperation, it is essential the firm maintains well-established, intensive connections with the users and that it follows the demand trends into the future.

Working along these lines will lead first of all to user-orientated basic concepts, then to promising prototypes and will finally result in those high-

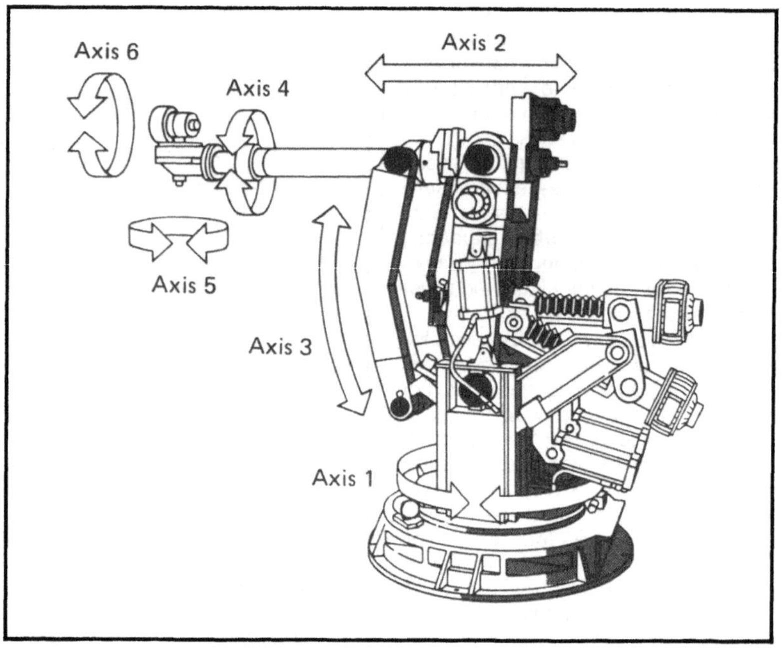

Fig. 1. The basic motions of the IR 601 machine.

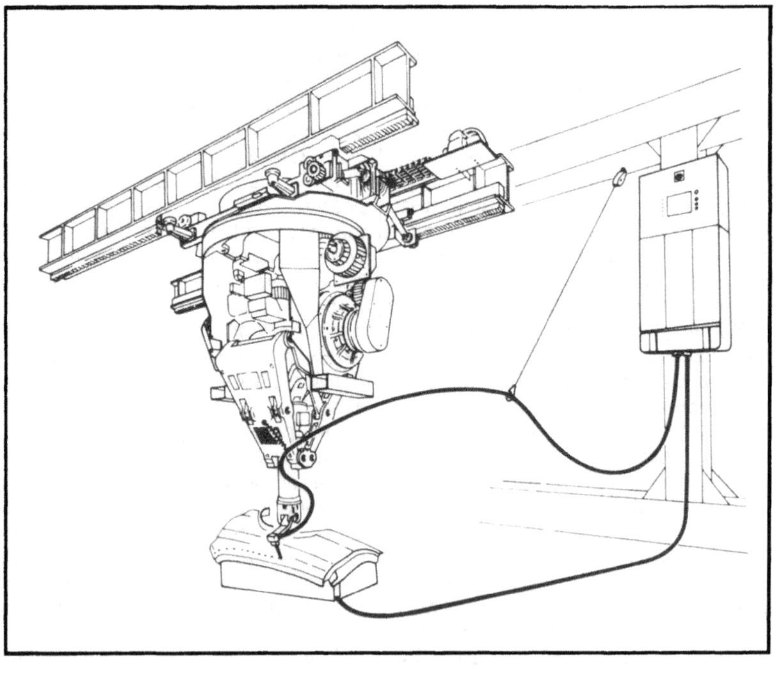

Fig. 2. IR 200 portal robot is designed for spot welding.

capacity series-produced systems which gain worldwide recognition as first-class achievements or as milestones of robot development.

What are the distinctive features of industrial robots which are said to be among the best in the world? As already mentioned, they must be robust and, despite their specialisation according to their particular application, they must be capable of performing the maximum possible variety of work.

The KUKA IR 601/60 floor-mounted robot, for example, is indeed used principally in the European automobile and agricultural machinery industries for spot welding with welding guns (Fig. 1). However, if fitted with the appropriate sensors, it is equally capable of arc welding, sorting and assembling.

Another KUKA robot, the IR 200 portal unit (Fig. 2), was primarily designed for spot welding with single pivoting thrusters or welding cylinders against backing electrodes, but provision was also made in the design for a broader range of application with the result that this robot can additionally perform drilling, milling and handling functions.

A top-class robot must also have a low maintenance requirement. Claims made in brochures in this respect do not mean a great deal; the experience of the user is all-important. The user's stringent demands are best met by far by all KUKA robots. This is to be attributed to the design concept, for example, the exclusive use of electric motors for drive purposes, the extensively modular design of the system elements and the possibility of rapidly exchanging a complete robot without the need for time-consuming programme corrections.

This enumeration of features of first-class robots should include also a low volume requirement along with a large working zone. The KUKA IR 601/60 is a good example in this respect and the IR 200 could hardly be bettered.

A high operating speed (up to 3.7 m/s in the case of KUKA robots), high load capacity or welding force (up to 100 and 500 daN respectively), good positional repeatability (to within ± 1mm) and low noise level (below 74 dB(A)) are further

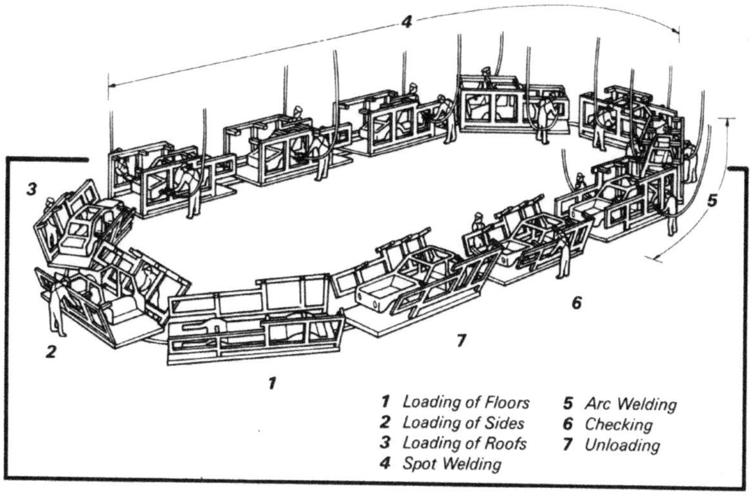

Fig. 3. Oval manual car body welding line.

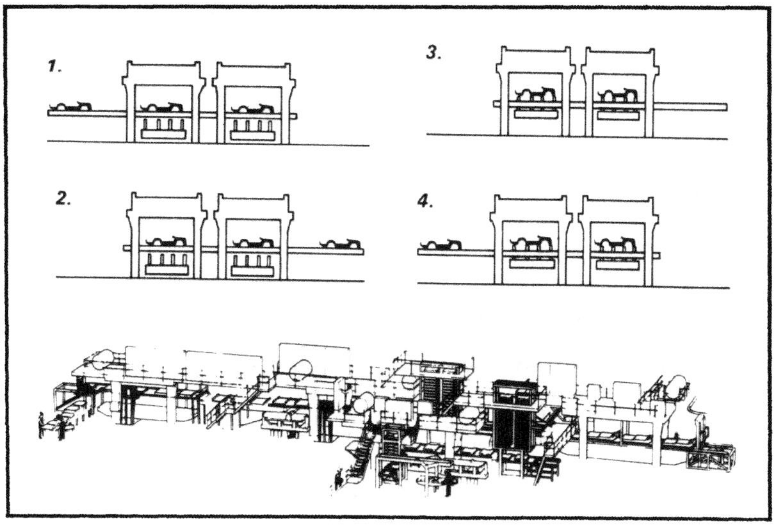

Fig. 4. The press-shuttle system transfer line.

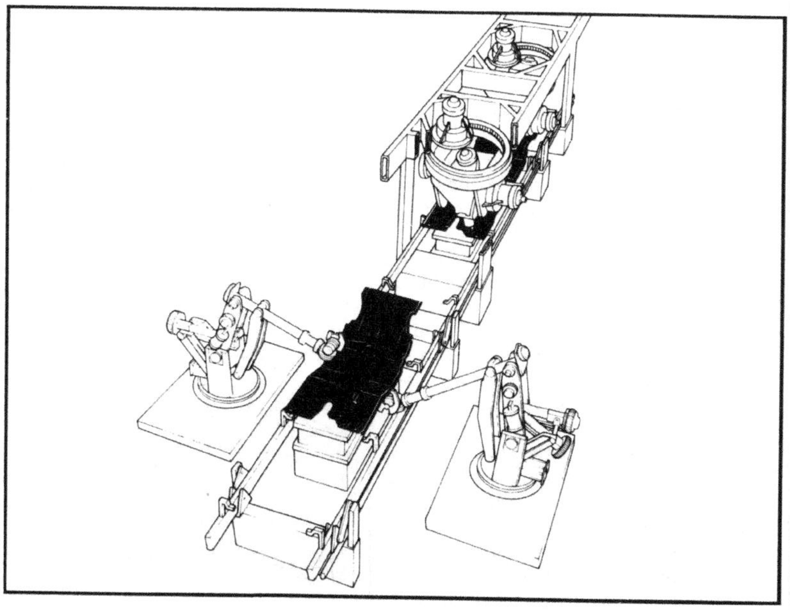

Fig. 5. The KUKA Autoflex system with IR 601, IR 250 and lifting shuttle.

characteristics displayed by successful robots.

Nowadays, spot welding robots are generally incorporated into flexible production systems for welding complex components. This has not always been the case, as the following example of the construction of car bodies in the automobile industry will demonstrate.

Up to 1960, car bodies were primarily assembled on oval assembly lines. The individual parts were brought to the main production line via conveyor systems. This was the first mechanisation of car-body construction. The output per shift was about 500 car bodies (Fig. 3).

'Since 1970 there has been increased use of FMS'

Since the 1960s, custom-made transfer lines have been used, employing presses with shuttle transfer. In these transfer-lines, the loading operations are triggered automatically and the spot welds are placed automatically and precisely (Fig. 4).

All the motions and welding operations carried out on the line, as well as the call-up and loading of the individual parts are monitored by a sequence control. These systems have a production rate of 800–1,000 parts per shift. Cycle times of 15s are possible.

Since 1970, there has been increased usage of flexible manufacturing systems with industrial robots; at first out of interest for a new means of production and then later because the application of flexible manufacturing techniques proved to be economically viable. So far, these new manufacturing systems have been primarily suitable for medium-range production rates – corresponding to 600–800 units per shift and thus slightly longer cycle times (Fig. 5).

The cycle time difference between transfer lines and flexible manufacturing systems with industrial robots requires a few words of explanation. In the case of robot-equipped systems with cycle times below 30s, the amount of time required for positioning the parts is relatively long. As a result, there is so

Fig. 6. Spot welding an automobile floor assembly.

little time left for spot welding that the robot is not used to its optimum capacity.

However, the automobile industry also wishes to benefit from the obvious advantages of fiexible systems for higher production rates. Consequently, the welding system manufacturers, in close cooperation with the automobile industry, continue to develop and improve overall system designs within which the spot welding rates are optimised and the robots operate at full capacity and thus economically.

At this point some remarks on the labour-related aspects of the use of robots are called for. The transition of the manufacturing method from the rigid transfer line to the flexible system, as outlined above, has no effect on the number of personnel basically required for operation and maintenance of the production line. Of course, the qualitative requirements demanded from these men are higher and this is where an old recipe proves its worth: training the existing operation and maintenance personnel in good time and thoroughly.

Statistics give an even clearer indication of the relationship between the application of robots and the number of employees. Between 1971 and 1981 the number of people employed in the West German automobile industry rose by approximately 12% while during the same period considerable investments were being made in advanced production systems, including industrial robots.

Moreover, this decade also saw a decline in the Federal Republic of Germany's annual automobile output from 4 million units to 3,900,000; the industry nevertheless managed to achieve growth – growth in terms of quality, finding expression in even more sophisticated, for example, flexible,

Fig. 7. Spot welding complete car bodies at BMW.

production systems and technically higher-grade vehicles. Growth, therefore, with industrial robots and at the same time an increase in the number of jobs.

This trend is of course not so pronounced in all industries. Robot advocates and opponents do, however, agree that industrial robots relieve personnel from unpleasant physical work so that they become available for monitoring and maintaining the new, fully flexible production systems. If in some instances the use of industrial robots reduces the number of jobs, this is offset by the newly created possibilities of working elsewhere, for example in the robot industry.

The labour-related discussion surrounding the use of industrial robots bears strong national characteristics, as cannot be expected otherwise in a free-market and autarkic economy. At the same time, however, the question of whether or not an industry should rationally be made more flexible with the aid of industrial robots is a European one. For example, a European automobile industry, with its host of ancillary industries, will be able to maintain or improve its position in the world market only if it offers its products on competitive terms. A prerequisite for this is the use of industrial robots. A flourishing automobile industry on the other hand will safeguard and even increase employment, as has been described.

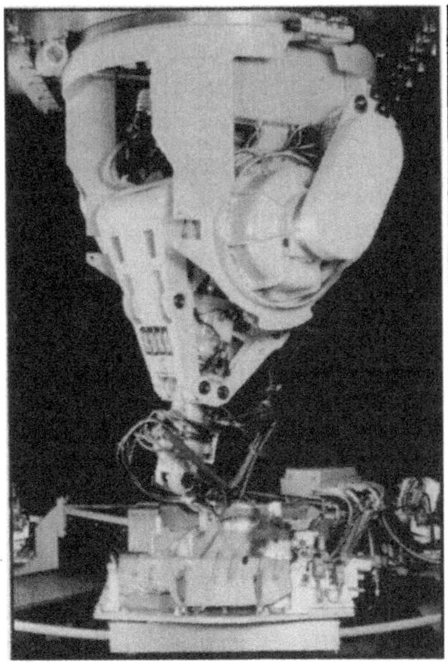

Fig. 9. IR 250 welding passenger car wheelhouse assemblies.

In the flexible automatic manufacture of car bodies, the welding equipment, such as welding guns, are moved by industrial robots. These flexible systems fulfil essential objectives set by the automobile manufacturers:

☐ Retooling production lines for different models or component alterations
☐ Better adaptation to capacity changes, i.e. to increase or reduce production rates
☐ Mixed production of different models on the same production line
☐ Greater manufacturing efficiency, such as by means of stand-by equipment.

'The robots reach the least accessible joints'

The economic efficiency of an automobile manufacturer and thus an important share of this manufacturer's competitiveness increases along with the degree to which these objectives are fulfilled. This is the reason for the automobile industry's demand for flexible automatic manufacturing systems.

It is easy to understand why this demand is gaining more and more importance, bearing in mind the increasing shortages of energy and raw material supplies, the rapid development of new techniques, the growing demand for large ranges of model variants and the increasing need for quick response to the demands of the market.

A few application examples for KUKA industrial robots are outlined below. Fig. 6 shows two KUKA IR 601/60 robots and one of the IR 200 series spot-welding a part of an automobile floor assembly. The two floor-mounted robots work with welding guns. The portal robot welds with a single pivoting thruster against a backing electrode positioned partly under the work piece.

The working station pictured here is part of a flexible transfer line equipped with conventional multi-point welding presses and a considerable number of industrial robots. Its flexibility consists largely on the ease with which model changes can be effected or mixed-type production carried out.

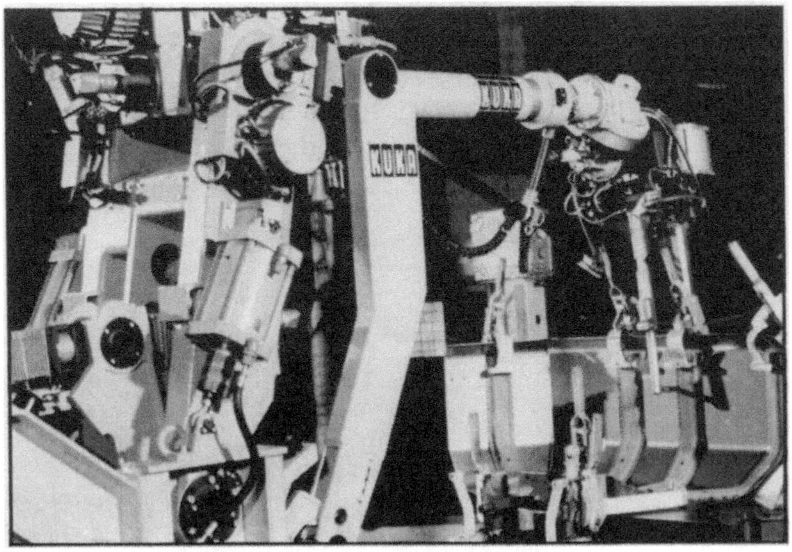

Fig. 8. Spot welding fenders and front hood parts of tractors with an IR 601/60.

Fig. 10. An IR 601/60CP mounting passenger car wheels using a visual sensor.

In Fig. 7 KUKA IR 601/60 robots are welding complete car bodies. Respotting car bodies on conventional transfer lines is no longer economically viable. Nowadays flexible, robot-equipped systems are used in the main, which adapt themselves perfectly to the frequent model changes and mixed-type production. The robots, which are equipped with special welding guns, reach the least accessible joints due to their flexibility and reach.

In Fig. 8 a KUKA IR 601/60 is welding fenders and front hood parts of tractors. Four manually operated devices are placed around the robot. Two of them are rotary and receive the individual parts of lefthand and righthand fenders which the robot then welds together.

At the two opposite movable devices, the robot connects the individual parts of front hood halves. The robot has no problem in manipulating the 90kg welding gun. The robot's travel speed is slightly reduced for this purpose. The required welding programs are initiated automatically.

Fig. 9 shows one of two KUKA robots of the 200 series. They weld left and righthand passenger car wheel house assemblies. A four-position turntable equipped with backing electrodes and clamps is used to locate the components for the placing of programmed welds. The turntable only makes a 90° turn so that each robot welds a lefthand and a righthand assembly alternately.

As demonstrated by these examples, KUKA has considerable experience with regard to the construction of industrial robots and also to the development and application of flexible manufacturing systems with such robots. The prerequisites for their use in car-body manufacture are the following:

- ☐ creation of reference points for all models
- ☐ robot-adapted car-body design
- ☐ assembly-adapted car parts
- ☐ accurately made stampings
- ☐ centralised production control
- ☐ qualified maintenance personnel.

The most important elements for flexible automatic car-body manufacture are the following:

- ☐ positioning devices for the welding tools, i.e. industrial robots
- ☐ welding tools, such as transformer guns, single pivoting thrusters and so on
- ☐ carrying and conveying fixtures, such as lifting shuttles, turntables
- ☐ data storage systems
- ☐ sequence controls.

'The result of long-standing co-operation'

All of these elements form part of KUKA's delivery programme. They too – like the robots themselves – are the result of longstanding close cooperation between the automobile industry and KUKA, both of whom are interested in finding optimum overall technical solutions.

The leading role of KUKA robots in the spot welding sector includes their outstanding suitability for use in other areas of application. The palletising of glass-tube bundles, the charging of annealing furnaces and the mounting of car doors and wheels (Fig. 10) are just a few examples of the extremely varied scope for using these robot systems. They also serve as indisputable evidence of the absolute top quality of KUKA industrial robots.

Arc welding: A difficult path for robots to tread

John Weston, principal research engineer at the Welding Institute, looks at some of the problems robot engineers have to overcome

THE use of robots for arc welding is now routine and there are many well established production installations. Worldwide manufacturers of machines suitable for this task have grown from three or four in the mid-1970's to more than 30 today.

Robots are available in many configurations but usually consist of a post mounted arm and wrist having five or six degrees of mechanical freedom, Fig. 1. For arc welding the welding head must be moved smoothly at relatively low velocities (8–25mm/sec), have good nett path repeatability ($\pm$0.5mm), and yet retain the capability to be moved rapidly between weld paths.

Welding robots are programmed using all of the standard techniques.

An important aspect of the arc welding robot system is the linked workpiece handling device, Fig. 2, which can add a further one to three degrees of freedom. A manipulator is desirable for several reasons:

- ☐ It improves safety by moving the work into and out of the robot working area for loading and unloading.
- ☐ It increases the work rate, for while one station is being unloaded/loaded the robot may be welding at another.
- ☐ It improves welding head access. Moving the workpiece increases the accessibility to the weld seams, permitting more welding to be carried out following one 'set-up' of the component(s).
- ☐ It increases welding rate. Specific joint attitudes can be welded at considerably higher rates than others, and a positioner permits the weld to be positioned in the most favourable attitude (e.g. flat for plate section fillets).

MIG or MAG arc welding equipment used in conjunction with robots usually comprises a thyristor controlled power source and a feed back controlled wire drive system. In general these units are combined and controlled in one or two ways.

'Selection box gives separate welding conditions'

The first, and at present most widely used, is via a selection box. This selection box has a number (three to ten) of 'sets' of voltage and wire feed rate control rheostats, each of which is adjusted manually to give separate welding conditions suitable for the components to be welded. Then during the path program the robot selects the appropriate 'set' of conditions.

In the alternative system the robot has a dual digital to analogue (D/A) output facility which is used directly to control, in proportion to this analogue output level, the rates of wire feed and levels of voltage.

Control through a selection box has the advantage of permitting rapid and easy adjustments to be made to welding conditions even while arcing is in progress. However, the rheostat levels are not part of the stored and recordable path programme and may not be recalled as repeatably as they are when D/A facilities are available.

The working tool of the arc welding system, the welding gun, is usually water cooled to permit high rates of working. Several techniques for the removal of spatter from the shielding nozzle are used either alone or in combination.

For example, systems which are operated as needed in the between weld periods include the pulsing of a strong air blast, often carrying an anti-spatter agent through the nozzle or activation of an internal sliding scraper to eject the spatter. Alternatively, the robot may move the nozzle to a station holding a rotating wire brush or reamer. The welding head units are essentially

Fig. 1. An ESAB/ASEA robot welding system at the WI.

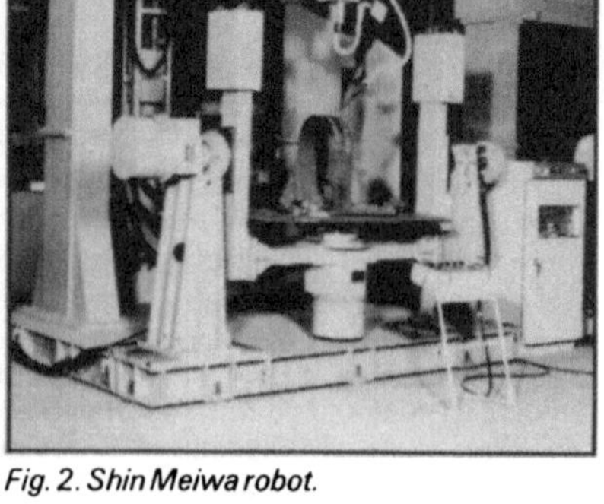

Fig. 2. Shin Meiwa robot.

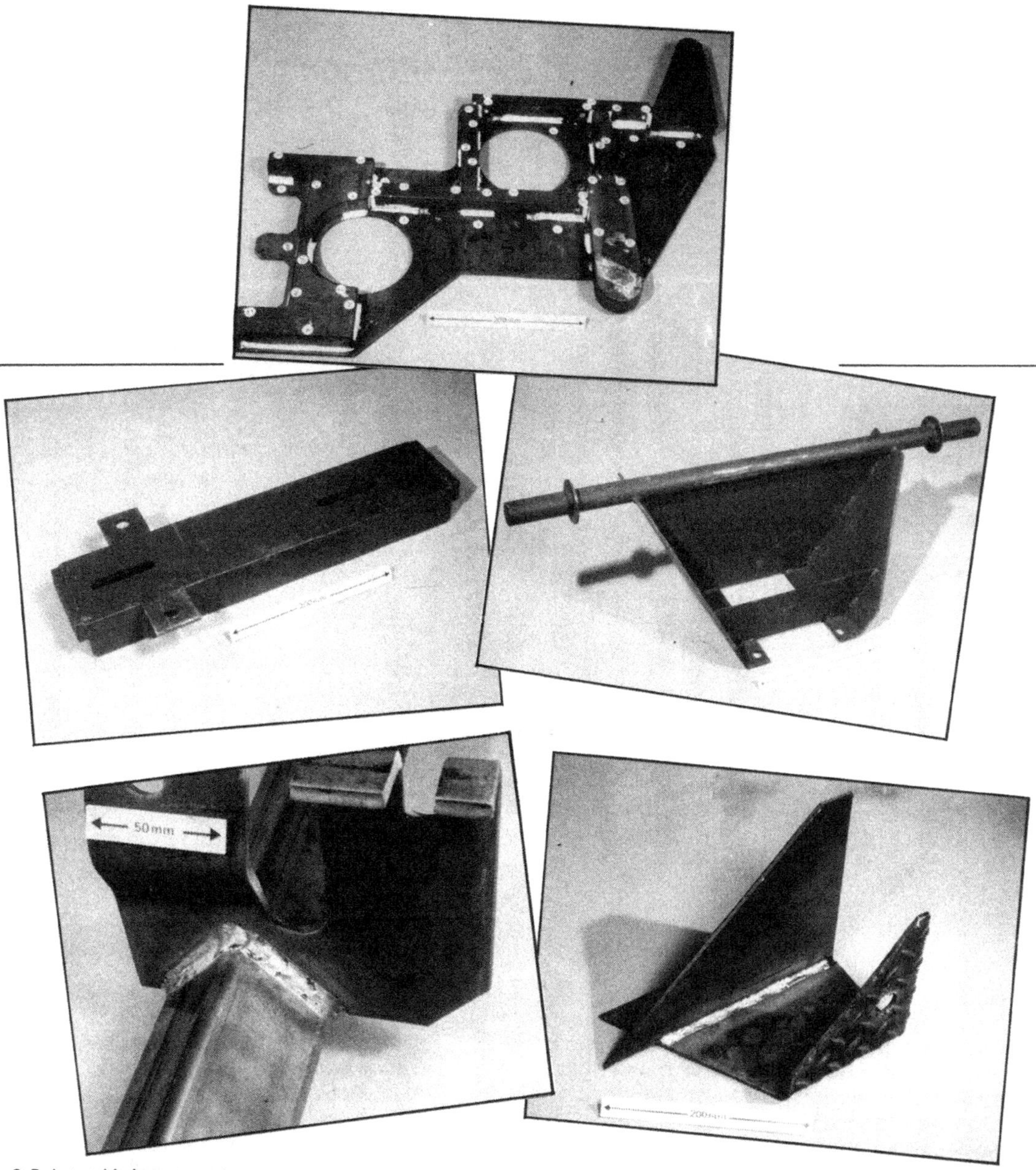

Fig. 3. Robot welded components.

modifications of semi-automatic MIG systems and suffer the problems common to these.

The welding consumable is pushed through a long length of conduit which is often heavy and stiff, as it also carries the services of air, gas, water and welding current. While it is difficult to envisage an ideal alternative some advantageous changes must be possible, especially when the load carrying capacity of robots is considered. There is clearly a need for further development, but this should not detract from the success of robot welders as they stand today.

The most important of the other arc welding techniques being used with robots is the TIG process. The basic welding equipment is readily controlled, as described for the MIG processes or, from units which are standard items within a TIG power supply.

The high voltage, often at high frequency, required to initiate the arm imposes special problems for electronic control systems which must be rigorously screened to eliminate interference. Where filler wires are used three degrees of freedom in the robot wrist become essential.

Case studies, Fig. 3, have highlighted several features of robot welding systems.

The first of these relates to the blind nature of the machine. The robot can be taught to reproduce accurately in space a sequence of weld paths, but in general it cannot see whether the component has been positioned accurately. Thus the importance of jigging and fixturing in locating weld paths repeatably with respect to the robot is clear.

It is those aspects of jig and fixture design related to the welding head access and attitude capabilities as determined by the robot that require the most forethought and skill. The aim must be to maximise the ratio of welding to non-welding movements within the limits of the robot and any associated workpiece positioner.

In considering the jigging system, consideration should always be given to the possibility of combining the tacking and welding operations so that the maximum productivity gains can be made.

The ability of the robot to interface with outside equipment can make it possible to open clamps which obscure a weld path. once tacking has been completed, or to add parts once sub-unit welds have been completed.

As well as repeatable jigging there will probably be a requirement for

higher accuracy and/or repeatability in the production of the components themselves. This may come from simple changes in the preparation cycle but could also require significant changes in the production route.

It has proved difficult to relate robot and manual cycle times because in general there are significant differences in:

□ the workpiece presentation
□ the welding task(s)
□ the learning and training times
□ the quality standards.

As already discussed, the component must be presented accurately and repeatably to the robot welder, which is seldom the case for the manual welder.

The robot will often weld more than one component at a time and can simultaneously weld a product mix, the work load can be such that the machine loader gains some of his relaxation within the robot work cycle.

'Cycle time takes no account of training'

Cycle times do not account for training or learning time. Having decided on a method of jigging, a process which in itself can involve a considerable time, the weld path must be programmed, tested and edited. This almost certainly takes longer than would be required by the skilled welder, and needs to be written off against, say annual production.

The quality standards required from the machine will almost certainly be different from those demanded of the production welder.

However, some comments can be made which permit a reasonable assessment of productivity to be made. The actual arc welding rate is essentially the same for the robot and for the man, and the gains come in the first instance as a result of the rapid between-weld movements, and the lack of delay at weld starts.

Furthermore, the robot dos not suffer from the effects of heat and fatigue as does the manual welder. It seems that on average robot welders can perform at between two and four times the rate of the human operator.

The robot cycle time is repeatable and fast, consisting of 50–80% arcing time. With fast cycles small time savings can make significant overall reductions and the largest gains can be made by reducing arcing time.

In making any such changes careful consideration must be given to the weld quality, the component design, and its ultimate production use. Design changes can have additional benefits in that they permit consideration to be given to modifications which ease the achievement of weld seam spatial repeatability.

Shortage of skilled personnel, rising labour costs and increasingly stringent environmental regulations are forcing the increased use of mechanised welding systems and robots. This move must be justified in comparison with present techniques.

It is a basic fact of welding production that the most widely used process, Manual Metal Arc (MMA) welding, requires very little capital equipment. It is a simple process and in machining technology may be compared to the use of the hacksaw and the file.

Even so the skilled operator can weld almost anything and can make adjustments to a wide range of fabrication environments but his productivity tends to be low and the quality of work produced variable.

Against this background arc welding mechanisation in any form, and via robot technology particularly, represents a big advance into the unknown for many organisations, and the first components to be studied should therefore be as simple as possible.

This approach permits maximum in-house experience to be gained and at the same time maximising the chance of attaining the economic goals forecast. Such experience permits other factors, usually listed as non-economic, e.g. improved quality, reduction of scrap, removal of undesirable jobs, to be couched in economic terms, thus leading to a more useful evaluation of the robot welding system.

Consideration of robot welders often forces a closer look at the whole fabrication system. The demand made by robot units upon other elements of the production line (plate preparation, component design, etc.) may be far reaching.

In spite of the present limitations of robot and welding systems, there are many tasks that present-day 'robots' can, and do, weld satisfactorily. These components range from the sheet steels used in automobile manufacture through to the heavier plate used in the construction of earthmoving equipment.

'Improvement in welding heads is needed'

Justification of the use of a robot welder in a manufacturing system will be an economic one, and this is often assessed on the increased productivity anticipated. As discussed earlier, 50% is a conservative estimate of the arc time in any robot working cycle. If the actual arcing time required to weld a component is known, doubling gives a conservative cycle time which can then be used to estimate the potential benefits.

There is a need for improvement in welding head and workpiece manipulators. The aim must be for smaller wrists with three degrees of freedom on the head manipulator and for more programme control of the positioners. Improvements are also desirable in the welding machinery as outlined earlier.

Software development is also required both to aid weld path programming and to give welding process control. Robot manufacturers recognise these problems and are making developments in these areas.

More sophisticated robot units equipped with sensory systems which would give 'sight' and permit joint recognition and seam following, would be more readily assimilated into production. Such machines should permit some relaxation of present jigging and plate preparation standards without loss of quality. They should also be easier to programme, which would reduce the economic batch size.

As confidence is gained by the welding industry continued increase in the use of robot arc welders can be anticipated.

Laser drilling: Standing on the threshold of technology

WE ARE on the threshold of an exciting new technology, the wedding of two separately identifiable, well developed but still maturing technologies – robotics and laser machining. Possibly the most significant area of application of this will be that of robot-aided laser drilling. But right now, robot-aided drilling with more conventional small tooling is the state-of-art technology.

Robert D. Bannister, Supervisor, Applications Engineering Industrial Robot Division Cincinnati Milacron Lebanon, Ohio, USA

Because of the many thousands of holes for fasteners in the typical aircraft, the use of computerised robots for drilling and related hole-making operations, for routing and for riveting is rapidly developing a broad base in the airframe industry. The production of these holes and the application of rivets and other fasteners have been highly labour-intensive and costly operations.

Workpiece size, frequent compound curvatures and other complexities, and the usual light gauges (largely sheet metal) are factors that discourage the use of machine tools, especially for hole-making operations. The use of portable power tools had become the standard. And the application of robots was inevitable.

Probably starting in 1978 on a developmental basis, robot-aided drilling is now moving into full production in various US airframe and tributary plants. And riveting is a close follower. Plants scattered through other industries, also faced with the problem of producing multitudes of holes, are likewise turning now to the robot as a tool for greater productivity and reliability.

Early developmental and production work in robotic drilling was largely restricted to use of Cincinnati Milacron computer controlled, hydraulic-drive robots. The electric-drive robots had not yet reached a sufficient level of sophistication. Since this is no longer the case, it is useful to look at the present and future for both.

The critical attributes for a robot in drilling and related applications include computer control with ample memory capacity; ability to interface with other computers, pc's and sensors; the range to reach as required by the size of workpieces encountered and possibly to reach tool racks, conveyors or other equipment; dexterity through multiple axes to produce holes over a large area, with each hole normal to the work surface, even on work of compound curvature; payload carrying capacity; point-to-point velocity; positioning accuracy under load at maximum reach and at full velocity; and end thrust capacity. These required attributes can be quantified according to the specific application.

Robots either present the tool to the work or present the work to the tool. Most experience to date has been with the former approach, where workstations or cells are the same or nearly the same as in past practice, but with the robot doing what an operator did previously. Here, the work is clamped against a template or drill-bushing plate in a jig or fixture that maintains part alignment relative to the robot.

The robot then moves a portable drilling unit from bushing to bushing to do a sequence of holes, each normal to the work surface, following computer commands. Many of the drill units used are self-feeding, although this feature is not necessary when using robots of sufficient capacity to feed the drill directly.

However, the dual use of self-feeding drill units and the robot permits a further step in productivity. Lock-on bushings are sometimes used, with one mating part of the bushing adaptor mounted on the drill unit. The robot positions the drill unit, the bushing adaptor parts engage and lock with a twisting action, the robot releases the drill unit and moves to pick up another drill unit and position it in place at another location, while the first unit is drilling.

The drill template of course is special-purpose tooling for a given part and its relative high cost must be considered in terms of the quantity of pieces required. The main body of the jig or fixture may be designed to accommodate a family of similar parts and corresponding interchangeable templates. The workpieces are often moved to the workstation by conveyor and are automatically or, usually,

Fig. 1. Drilling holes in an aircraft fuselage with a T^3.

manually loaded into the jig or fixture. In future systems, the robot will frequently automate this function.

In many applications, the hydraulic-driven robot offers sufficient positioning accuracy since the hard tooling incorporating tapered drill bushings provides the final accuracy required. The amount of compliance inherent in the hydraulic drive system is usually sufficient, but occasionally further compliance must be built into the tooling.

'Electric drives offer greater accuracy'

With the more recent advent of highly sophisticated electric-drive robots having the configuration and other attributes to participate in this broad area of application, other possibilities have emerged.

The electric-drive robot has greater positioning accuracy and repeatability. It has become the standard for high-tolerance work where the accuracy and repeatability of the hydraulic-drive robot were doubtful. Its lack of hydraulic-drive compliance is an asset to maintaining accuracy but in some cases requires compliance in the add-on tooling or in some cases requires greater dependency on sensors.

However, the accuracy of the electric-drive robot will often make unnecessary the costly drill templates, permit even more flexible automated cells, and provide the economies of automating short-run and even prototype work.

The second approach to robot-aided drilling is that of presenting the work to the tool. In these applications, the drill unit (or units) is mounted permanently or semi-permanently in fixed position and the robot lifts the workpiece from the conveyor and presents it in a series of positions and attitudes to the tool. Without need for drill bushings, this approach will be used more frequently. This approach has also been used with riveting where the riveting machine was quite large.

The Cincinnati Milacron electric-drive robots have some other advantages. They have a unique three-roll wrist which gives the robot additional dexterity, even to pointing back at itself. This will often enable the robot, after drilling a panel and changing tools, to reach the backside of the panel to deburr the breakout end of the holes where required.

Fig. 2. Drilling a curved canopy.

In fact, while we are presently seeing some combination of operations in a single work cell, in the future it should be more common to see a series of operations required on the workpiece done in one setting of the workpiece.

On very large workpieces, a positioning table or conveyor may be necessary to index the workpiece or to traverse it linearly past the robot's location in order to make the holes in subsets. Or, the robot can be mounted on a track to travel from position to position parallel to the workpiece.

In the past, the work was usually held vertically so the drill operator could stand in front of it and drill horizontally (with heavier drill units sometimes being partially supported and suspended in a counterbalanced sling). Today, process planners have much more flexibility, the robot can even be mounted to the ceiling if desired.

Many of these comments apply equally to riveting. To date there has not been so much cumulative experience with robot-aided riveting, but that experience indicates great promise for such applications. The primary considerations are the weight of the tooling and the ability of using opposed-force tooling. Parts to be riveted are held together mechanically or by adhesive tape as in normal practice.

Today, on some subassemblies, parts can be routed and drilled and riveted in one setting of the work. The robot simply has all the tools it will need in a readily accessible rack and under programmed command automatically moves to the rack, sets down the tool just used and picks up the next tool needed. This automatic interchange of tools and end effectors by the robot has become a well accepted art.

Today's robots are more intelligent and with state-of-art sensors and others just down the road, they are broadening the field of application for robot-aided drilling, riveting and related processes. If required, as when impact from an external source is detected or on very large workpieces where precise locating is difficult, the robot can be programmed to automatically and quickly go to a reference point in three different axes to assure part orientation and to shift its program to offset any small error in part location in the X, Y and Z directions.

Drill wear in high-production work has been a problem. Sensors are now available that, as on machine tools, can continually monitor the drilling force and compare it with a reference value, making a severity level decision and signalling the operator or stopping the operation when the force exceeds a given limit indicating a worn or broken tool. Similarly, on riveting operations, force sensors can protect against driving a rivet into a spot where a hole for some reason does not exist.

There still remains need for better solutions to several problems, chiefly in terms of sensory feedback. We need better means to detect and measure changes in the working environment, wear on the drill bit, broken drill bit, workpiece out of alignment, and other conditions and to react appropriately.

We need also improved vision systems to identify workpieces received in random order, to verify their orientation, and to detect abnormalities or to

find certain workpiece features that serve as reference points. When riveting we need to be able to more assuredly locate hole positions and determine whether they were drilled or not drilled. All of these problems are being attacked by extensive R & D effort and are close to being solved.

Laser technology has now advanced to where the robot-aided drilling of various materials by laser offers considerable potential. This application was demonstrated at IMTS-82. Aside from the speed with which a laser can drill a hole, advantages include these: there are no consumables, no reactionary pressures on robot or tools, no tool deflection or breakage, no tool wear or tool sharpening, no burring, no chip removal problems, no drill templates. Since the laser units are relatively large, the robot will usually present the work to the laser.

'The more powerful industrial CO_2 lasers will have broader application'

Two types of lasers lend themselves to such applications. These are the CO_2-type laser and the Yag-type laser, a solid state device. A Yag laser produces high peak power pulse permitting drilling to occur very rapidly. Unfortunately, the Yag laser has not yet developed past approximately 100W or so of power and therefore these units are presently limited to small diameter holes and light-gauge materials.

In the near future, the more powerful industrial CO_2 lasers will have much broader application. One limitation is the factor of reflectivity; aluminium for example is not a good candidate for this process.

In conclusion, while the technological problems of robot-aided drilling and riveting have been sufficiently resolved to allow many production applications, there are still further developments in sight that will greatly expand the potential in this field.

The creeping technology of grinding and fettling

W. McCormack and E. G. Godding of the BCIRA examine the technology

THE use of robots throughout the world has been growing rapidly, but their lack of rigidity under reactive loading conditions such as those which occur in metal removal operations by grinding, chipping and mechanical cutting, as opposed to gas or flame cutting, has hindered the application of robots to the fettling of many types of castings.

In applications in which the robot picks up and handles a fixed load such as in paint spraying, weld or assembly operations, positional accuracy at the end of the robot arm is typically ± 0.4mm. Under reactive loads the arm may deflect as much as 12mm when forces within the rated load capacity of the arm are applied.

Success achieved in the use of robots for the fettling of castings other than those made from steel for which flame cutting may be applicable, has resulted from the identification of special problems of reactive loadings and the development of technologies at BCIRA by which they can be overcome.

'Robots can extend cycle times'

A second factor hindering the use of robots in fettling has been the assumption that methods of working, tools used and sequences of operations carried out in manual fettling operations merely have to be transferred to a robot and the required results will be obtained. This is a misconception. Using a robot to reproduce manual operations can extend cycle times to a point at which the use of a robot is not economic, whereas if a new approach is sought, the application of robots becomes attractive.

Castings too heavy for a man to manipulate can be picked up and handled easily by a robot. Likewise, tools of much greater power and weight can be used. In fettling with robots a 35kW 760mm diameter cutting disc and smaller 12kW hydraulically driven discs (Fig. 1) are regularly used.

In manual operations 2 to 3kW tools would normally be used. A robot handling a 3.7kW pneumatic tool to use burrs between 3 and 20 mmdiameter is shown in Fig. 2. In a foundry a robot is in regular use to remove flash in a hole using a 38mm diameter reamer. This is made possible by the slow precise feed of the tool into the hole which is possible using the robot and could not be carried out manually.

A tool far too cumbersome for man to handle is the abrasive belt grinder shown in Fig. 3. With robots such tools are easily used to obtain the advantages of much higher stock removal rates and often of accessibility.

The concept is emerging of a robot as an integral part of a fettling station which includes facilities for the delivery and presentation and removal of castings after fettling.

Different tools and casting handling devices can be rapidly interchanged using universal gripping systems containing their own power supply connections and controlled by the robot program. To obtain full advantage methods of casting production may need to be modified to ensure that metal to be removed is in a position as convenient as possible for the robot to reach.

Considerable progress has been made at BCIRA in the development of the use of robots for the fettling of castings which include domestic boilers, automobile engine blocks, valve bodies and gear boxes.

The techniques used which are either for the robot to pick up the castings and to apply them to fixed tools, or for the robot to handle the tools and to apply them to castings, are determined by the weight of the castings and by the positions at which fettling is to be carried out. In some instances both techniques may be used in the fettling of a particular type of casting to obtain efficient robot utilisation. The applications of both are described in the following sections of this paper.

The use of a robot to manipulate a casting rather than the fettling tools has the following advantages which enable a robot, with minimum additional handling equipment, to carry out the complete fettling of large numbers of similar castings without interruption, without operator intervention and perhaps without operator attendance.

Fig. 1. 300mm dia. 12kW abrasive disc.

- □ Castings are picked up from a fixed point, which can be magazine fed, from predetermined positions, su as from pallets having casting holdi fixtures.
- □ Without stopping for tool changes robot moves the casting to a succ sion of different tools to carry out required operations.
- □ High power tools can be used obtain the required finish in minimum time.
- □ If necessary, castings can be plac on to an intermediate fixture, turr and regripped to provide access further surfaces.

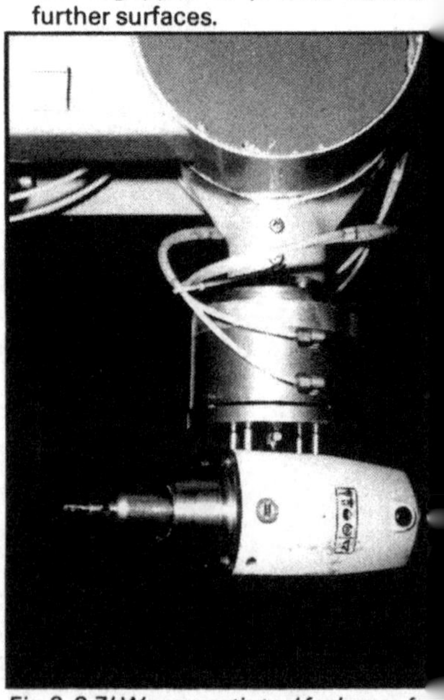

Fig. 2. 3.7kW pneumatic tool for burrs of 3-20mm dia.

Fig. 3. Abrasive belt grinder modified for robot use.

□ On completion of fettling, the robot can put the casting in a fixture, place it in a bin or on pallets.

Provided that the weight of castings plus grippers are within the limits of the robot's lifting capacity the advantage listed above often make this the preferred method for fettling operations. It extends the advantages of fixed tool operation to castings which would be too heavy to fettle in this way by manual means.

Fig. 4. Gripper for holding pan support castings.

The first step in the development of fettling operations in which the casting is held by the robot is the provision of gripping devices which positively locate the casting, hold it firmly during the metal cutting and removal operations and which desirably do not restrict access to any of the surfaces which are to be fettled.

'Grippers should be versatile'

The provision of individual grippers for a large number of different types of castings can be expensive. Therefore, grippers should be designed to be as versatile as possible to enable each to handle a range of components. Examples of grippers developed at BCIRA are shown in Figs. 4 to 6.

The one shown in Fig. 4 has been designed to hold the casting either internally or externally to allow both the inside and outside edges of the castings to be fettled.

The second gripper shown in Fig. 5 has been designed to hold relatively heavy castings. It embodies a power operated chuck and can accommodate castings of various sizes held by operation of the gripper either in an internal hole or on an external diameter. The gripper shown in Fig. 6 has been designed to hold a range of valve bodies which differ externally but which have the same throat dimensions.

The first step in many complete fettling operations is likely to be the removal from the castings of runners or risers. In the fettling of iron and non-ferrous castings, non-contact flame cutting is seldom applicable. However, abrasive cut off discs are frequently used.

Special techniques have been developed for the use of the discs with robots, since although the wheels are reinforced they can be broken by any lateral or twisting movement of the casting relative to the wheel, particularly when the cut is well advanced.

Under such conditions even if breakage does not occur the wheel will bind within the cut leading to excessive power consumption and wheel wear. A solution is obtained by the use of a

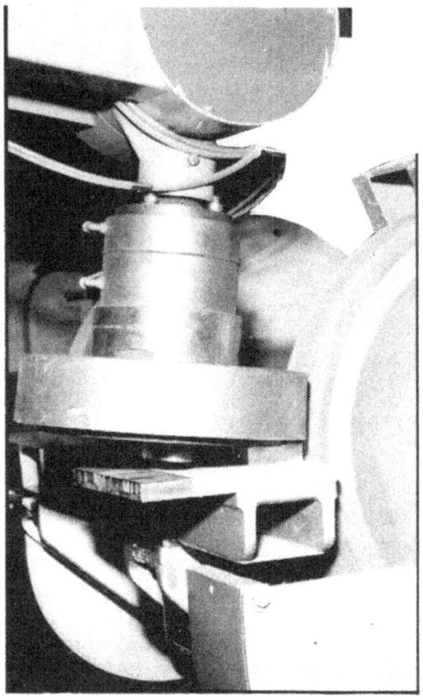

Fig. 5. Chuck gripper for circular castings.

wheel guidance system either attached to the gripper or in the form of a circular plate mounted at the wrist of the robot with guidance being obtained using a caliper attached to a sliding mechanism. The latter is preferred since attachments to the gripper can restrict access and more than one cut can be made in the same casting.

The final fettling operation on most

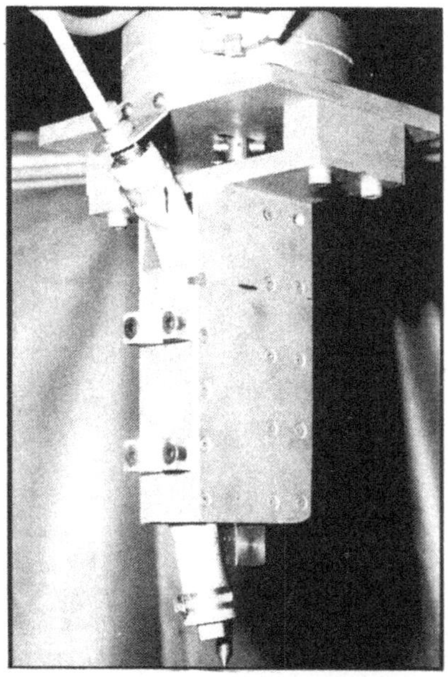

Fig. 6. Internal gripper for valve body castings.

Fig. 7. Abrasive belt grinding.

castings is carried out by grinding, although burrs may also be used. Such operations produce considerable reactive loads. Special techniques have been developed at BCIRA to overcome the inherent inaccuracy of robots under these varying reactive loading conditions. In addition, when grinding wheels are used techniques are required to compensate for wheel wear.

When heavy grinding operations are required it may be difficult to determine when sufficient grinding has taken place. Using adaptive control systems it is possible to sense when an increase in grinding pressure occurs and to compensate automatically within the robot program to offset this effect, but it is often quicker and easier to use an external guide template to control the movement of the robot. Pressure can then be applied in excess of that likely to be required and the casting will be ground to size irrespective of the amount of stock to be removed.

Where flash is to be removed or if the area to be ground is small compared with the surrounding area, the casting itself can limit the grinding action and special procedures may not then be required.

Wheel wear can be accommodated in one of two ways either to advance the tool in proportion to the amount of wear which has taken place or to modify the robot program. In the former case sensors can be used to detect the edge of the wheel and hence to drive the unit forward by the required amount.

Fig. 8. Form grinding using an abrasive belt.

'Belt grinders save wheel wear'

Alternatively, the same type of sensing mechanism may be used to modify the robot program, replacing an existing program by a new program for each predetermined increment in wheel wear. Spring loading of the tool can help, some small amounts of wear can be accommodated by compliance of the tool mounting allowing proportionally greater wear to occur before the robot program has to be altered.

The best solution to wheel wear, which also provides the advantage of increased rate of stock removal lies in the use of abrasive belt grinders. With these tools the cutting surface does not change its position during the life of the belt and hence no provision has to be made to compensate for wear. However, in order to obtain the highest rates of stock removal it is essential that the correct backing and work wheel is chosen.

Examples of the use of abrasive belts with a robot are shown in Figs. 7 and 8. In the first example use is made of the retention of a fixed cutting surface. In the second, the feature illustrated is the use of a selected diameter of contact wheel to match the contour of the casting section to be fettled.

For cutting and metal removal operations the same principles apply as in situations in which the casting is held by the robot. Again it is desirable that tools of maximum power should be used and that the choice should not be constrained by the availability of portable tools for manual operations.

It is unlikely that any casting can be fettled using only a single tool. It is therefore essential that the robot should be provided with a number of tools which can be selected and picked up by the robot as required. Tool change systems are needed which meet the following requirements:

- □ The system must be physically small so that there is minimum restriction of access to surfaces to be fettled.
- □ The tool must be positioned as close as possible to the wrist of the robot to avoid strain on this joint.
- □ Clamping systems must hold tools firmly and be suitable for a range of different tools.

Fig. 9. Universal gripper attached to robot arm.

Although clamping devices are available, most do not meet all of these requirements. BCIRA has developed a universal gripping device* for use particularly in fettling operations, although it is suitable in many other applications. This device is shown in Fig. 9.

It consists of a short stroke air cylinder operating a T-bar which engages a double key hole slot in an adapter plate (Fig. 10) mounted on the tool. The rotation of the robot wrist is

* Patent applied for.

used to rotate the T-bar when it has passed through the key hole. Positive accurate location of the tool is provided by a contact stub and two cones mounted on the cylinder which correspond with similar locating points on the adapter plate fitted to the tool. Power supplies to the tools can be provided by plug and socket mechanisms incorporated in the gripper.

A small number of robots are already in use in foundries for fettling operations and a considerable expansion can be expected to increase productivity and to meet social and regulatory pressures to improve working conditions by reducing exposure of operators to noise, dust and vibration. Existing technology enables much progress to be made, but further developments are inevitable.

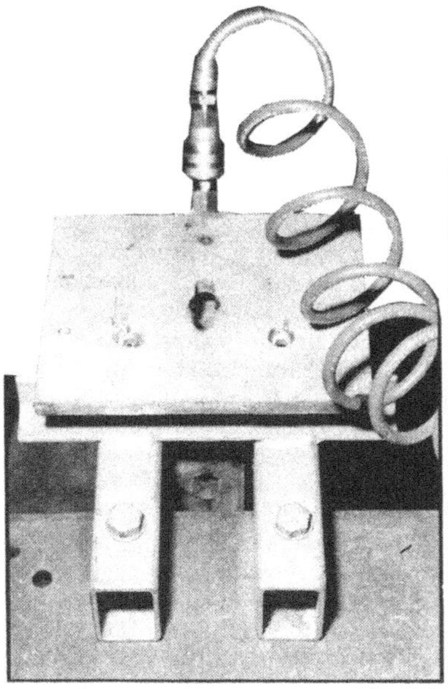

Fig. 10. Tool adapter plate for use with universal gripper.

These developments are most likely to lie in the fields of robot operation by the use of vision systems for recognising castings, picking them up from random locations and correctly positioning them for fettling operations, and by new systems for the control of robot movements having improved capability for sensing positions and amounts of metal to be removed and hence obtaining optimum rates of operation.

In the short term, further developments in tooling and techniques of handling castings will increase the variety of castings which can be handled by a single robot and will widen further the application of robots for fettling.

Saving the operator from a sticky environment

Mike Kelly, manager – manufacturing automation at BL Technology, shows how robots help stick car bodies together

THE use of adhesives in automotive body manufacture is a well known technology, but the application of these sticky and sometimes pungent materials is one of the more unpleasant manual tasks in the body manufacturing area. Not only does the operator have to reach at full stretch to apply the adhesive to the extremities of the larger panels, but must also move around the panel at a consistent rate to ensure the specified, regular bead.

The adhesive must be applied accurately to the defined locations in the specified amount, to ensure consistent quality. The operator is close to the adhesive and with some formulations the uncured adhesive can cause skin irritation, therefore manual application techniques restrict the range of materials which can be used.

'Applying adhesives offers little job satisfaction'

The application of adhesives is a monotonous, tedious task which gives little job satisfaction. Consequentially automation of adhesive application is beneficial to the operators and releases them for more rewarding work. It also allows a wider choice of bonding materials to be used.

Due to styling changes and market demands, vehicles, and consequently panels and assemblies undergo successive changes. Automated systems must be capable of adaption to meet these new requirements.

Dedicated automation is both complex and costly to change and a major modification will give rise to lost production time. Robot based systems can be quickly reprogrammed to meet the new situation without reducing production. A model or styling change will usually involve altering the position or size of the adhesive bead or spot or in some instances adding additional amounts of adhesive to the panel.

A robot can be programmed for many different panels and operate on them in any sequence or on a batch production basis. Such high flexibility makes an installation in a low volume environment cost effective.

British Leyland has robotised two previously manual installations on a retrofit basis. The first installation was the application of 63 individual 1ml spots of adhesive and a 3.5m 3mm diameter bead of adhesive to the trunk lid inner assembly for the Triumph Acclaim.

This component, at current production rates, required the operator to complete each assembly in less than 60s, a condition which could not be achieved without additional labour. The repeated operation of the manual applicator (2,450 operations an hour) gave rise to wrist strain and operator complaint.

The robot system installed was based on a Unimate Puma 550 series robot mounted in the upright position adjacent to the panel. Due to the panel geometry and the restricted reach of the robot it was necessary to fit a 300mm extension arm to the robot wrist flange with an automatic applicator mounted at the opposite end at approximately 45° to the wrist axis.

With this configuration the robot was able to reach the whole panel and maintain the applicator at approximately 90° to the panel surface during application. The trunk lid inner assembly is manually loaded into a simple jig fixture and locked into position before the cycle is initiated.

To give a clean cut-off to the adhesive flow a 'positive shut-off snuff-back' applicator valve was used. The nozzle orifice was adjusted to give optimum performance. The robot applicator is fed from a single double-acting 50:1 transfer pump giving a delivery pressure between 100 and 110 bar.

Using this high pressure enabled short times to be programmed for each spot application, and relatively high speed movement for the bead application. This resulted in a cycle time of 58s.

The second system is based on an ASEA IRb6 robot and applies a 4.5m 3mm diameter bead of adhesive around the periphery of the Mini Metro taildoor.

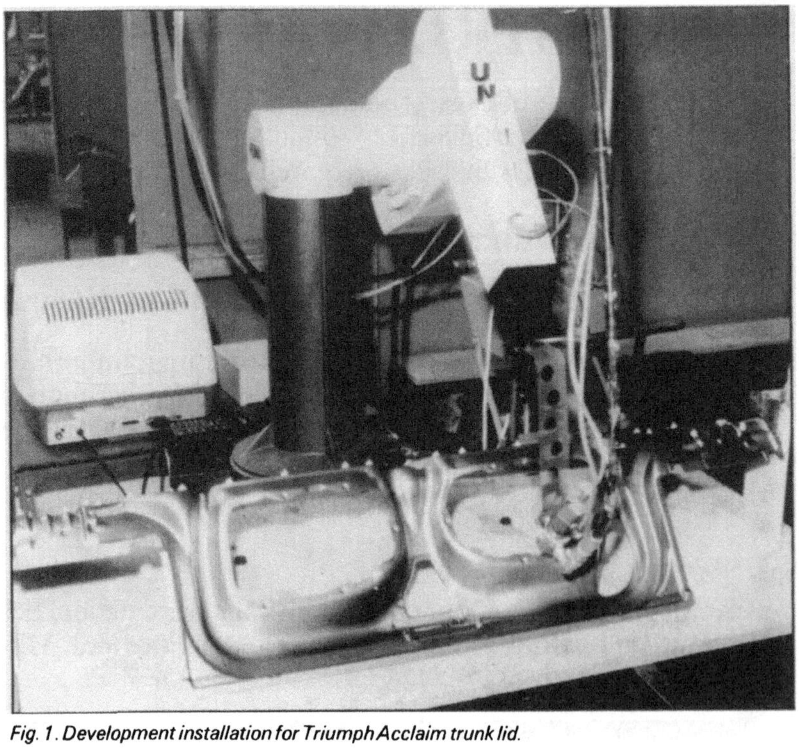

Fig. 1. Development installation for Triumph Acclaim trunk lid.

The production rate of 176 units an hour required that the adhesive cycle was completed in 12s, a rate which was matched by the robot.

Due to space constraints it was necessary to mount the robot inverted over the assembly track. Again due to the panel shape, size and the limited reach of the robot it was necessary to fit a 300mm extension arm with the applicator mounted at approximately 30° to the wrist flange. At this angle the applicator maintained an attitude of approximately 90° to the panel surface during application.

The outer panel is automatically transferred into the bonding station and locked into position prior to the automatic initiation of the robot cycle. The robot is fully interlinked with the automated Taildoor Assembly Line which produces a complete fully cured assembly for subsequent processing.

The adhesive is fed via a manifold system supplying both the robot and manual applicators from a twin pump auto-barrel changeover transfer system. The delivery pressure is typically 100 to 110 bar. Material surge as the valve is opened cannot be tolerated with this assembly and a pressure regulator is fitted close to the robot. This reduces the pressure over the final three metres to the applicator.

Both installations were 'robotised' after the initial model launch and had to be compatible with the existing facilities.

Manual application allows a fairly wide choice of materials to be used. In a robotic installation materials selection is important to achieve optimum application performance.

Robotic installations require that material is consistent in terms of cleanliness and viscosity over a wide temperature range. Changes in viscosity affect the flow rate and unless the robot is equipped with interactive flow rate detection, varying quantities will be dispensed.

BL specify the viscosity and cleanliness levels to be attained and monitor the supplier's manufacturing facilities and procedures to ensure that these requirements are met.

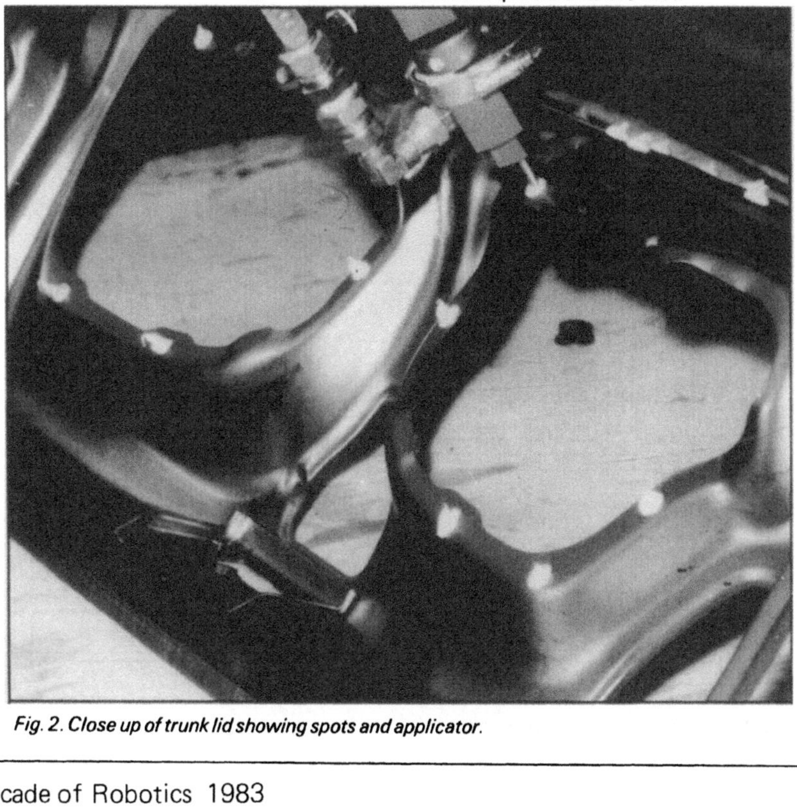

Fig. 2. Close up of trunk lid showing spots and applicator.

Both installations discussed have been retrofitted into existing facilities and therefore had to be acceptable not only to the plant management in terms of operating efficiency and cost effectiveness but also to the work-force in the immediate environment.

At each stage of planning, installation and commissioning personnel were kept fully informed of the aims, objectives and achievements. This was a two way exercise with the workforce actively participating. No problems were encountered.

The two robots are in different plants and as there was no relevant experience, training programmes were run for maintenance and process engineering departments. This was carried through the commissioning and pilot production period.

'Predicted manpower savings have been achieved'

In the case of the spot application facility the only preventive maintenance required has been the replacement of the applicator, seals and needles, this relates to 350,000 valve operations.

The applicator is pneumatically operated and slow compared to the robot commands, consequently it was necessary to build a short delay into the program to avoid 'stringing' as the robot moved from one site to another. The reaction of the valve is also influenced by the rheology of the material, the delivery pressure to the valve, the air pressure to the valve and the initial assembly of the valve itself.

With high delivery rates required from the applicator relatively large bore high pressure feed pipes are required. However these can apply undue strain to the robot wrist, unless counterbalanced, and affect the positional accuracy of the robot.

Accurate panel jigging is essential and with both installations the panels are located to within 0.5mm of the nominal position, in all directions.

These installations have achieved all the predicted efficiencies manpower savings and quality levels.

Machining cells put robots in their right place

G. Martins, R. Nordström and M. Svensson all of ASEA's industrial robot division in Sweden, look at robots to serve machine tools

MACHINE group serving by means of an industrial robot is an economic and effective method of optimizing the operation of NC-machines and other machine tools and production equipment of different kinds. This type of workshop equipment is an advantageous alternative to expensive specialised machines. As such machines are difficult to adapt to other purposes, their use requires an extremely specialised production of a long series to be economically defensible.

Robotised serving of machine groups is economically rewarding with even short production series as the machine groups, by reprogramming, can be quickly adapted a wide range of other tasks and working methods at any time.

In serving a machine group, the robot functions as material handling equipment and ensures the optimal utilisation of the different machines in the group. The robot can also take active part in the production itself by performing certain processes, for example, deburring or by serving measurement stations, cleaning stations etc. which can also be part of the group. The robot is an integral part of the machine group and its versatility can make possible the use of simpler machines in the group than would be otherwise necessary with other production methods.

'The group is controlled by a robot'

In comparison with traditional production methods, robotised machine group serving has the following advantages:

- ☐ Better utilisation of both machines and robot
- ☐ Shorter through-put time
- ☐ Improved and more even product quality
- ☐ Increased operational safety
- ☐ Lower personnel training costs
- ☐ Partly or completely unmanned production

Experience of production techniques is necessary for the design of an effective machine group.

A machine group is built up in accordance with the requirements of the customer for manufacturing a certain range of products. The manufacturing methods are often already established.

A machine group is then built up with the help of existing or newly acquired machines. These are modified for automatic operation and control from the robot. The system usually includes measurement fixtures for checking the details at different stages, equipment for cleaning, occasionally deburring, turn-tables, and so on as in Fig. 1.

To permit the machine group to work continuously during a certain time, a system is required for the supply and un-loading of parts. This can be a conveyor system or pallets.

The complete machine group is controlled by a robot system which starts, stops and checks all of the equipment in the group via inputs and outputs.

The robot can communicate with the peripheral equipment through a number of inputs and outputs, for example to start and stop machines and peripheral equipment. The surrounding equipment can control the execution of the robot program by means of input signals.

The signals between the robot and the other equipment are connected at the central panel. From there, the operator can start, stop and check all the group equipment and machines including the robot. The different safety functions are also placed in the central panel.

The following description refers to the example shown in Fig. 2 and Fig. 3. The input magazine holds items sufficient for production during one shift. The machine group can thus operate unmanned for a complete shift. A total of approximately 20 different items can be processed by the machine group.

The robot extracts the finish-machined item from the lathe chuck and inserts the next item with the aid of a double gripper. The lathe is furnished with a measurement control facility which enables the robot to check machining results.

After machining one side, the robot places the workpiece in a turning device which rotates the work piece 180°. The robot then re-inserts the workpiece in the lathe chuck.

When finishing is completed, the robot delivers it to a grinding machine for surface grinding. The workpiece is then drilled in one of the two NC-drilling machines after which it is placed on a pallet located on the outgoing conveyor.

By exchanging signals with the processing machines and gauges, the robot can check that items being produced are of the defined quality.

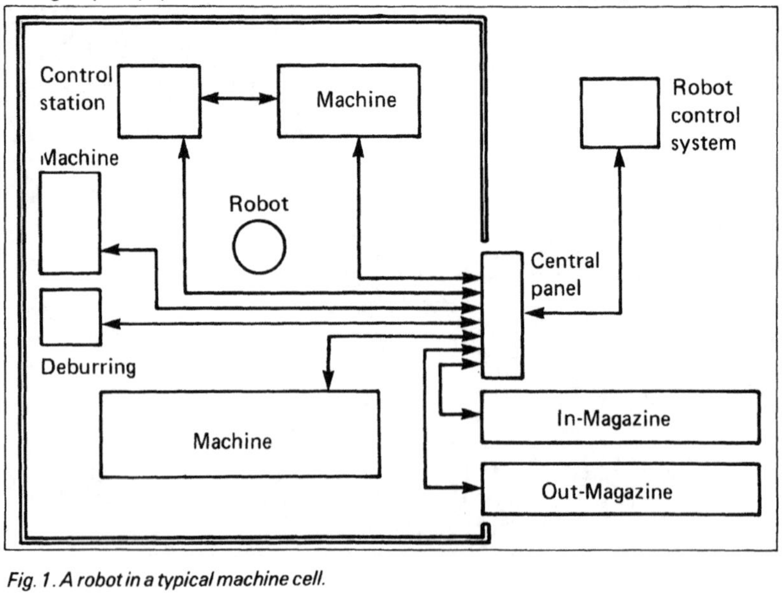

Fig. 1. A robot in a typical machine cell.

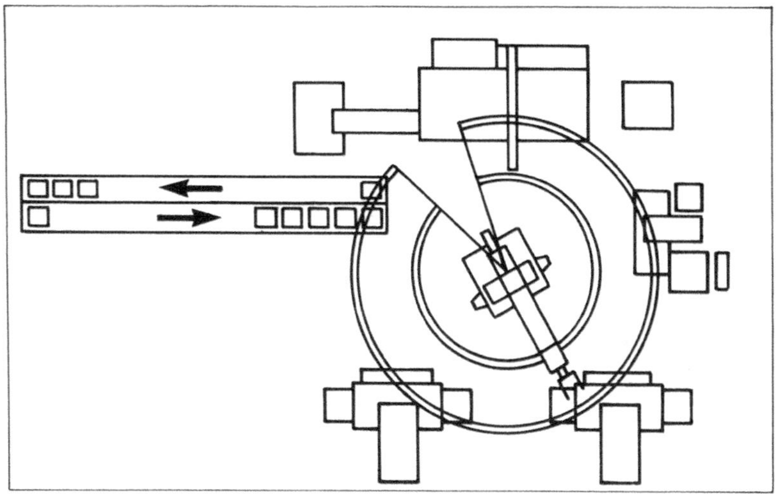

Fig. 2. A station layout for a robot-served machine group.

Fig. 2 An example of a station layout for a robot-served machine group

The preliminary design procedure includes an evaluation of the economic and technical feasibility of production automation with the machine group. This is based on the description of the existing production as described by the customer.

'Machines are more expensive than the robot'

The technical evaluation which then follows is intended to indicate if or not the proposed installation is feasible and if it is reliable. For this it is required that the designer prepares a layout drawing of the machines and the robot.

The programming procedure is based on the flow of material and details which are to be processed by the machine group and the sequence in which the processing is to be performed. The relative speeds of processing in the different machines must be taken into consideration.

The machines in the group are most often more expensive than the robot and it is therefore of importance that these can be utilised fully. A flow diagram can be prepared after an analysis of the flow through the group and the programming is based on this.

To reduce the risk of accidents, the machine group should be enclosed within a guard rail provided with a safety gate. When this is opened, the robot should stop automatically while the other machines continue their automatic cycles.

This technology started with cells designed for one, or at least a few objects with re-setups needed between batches normally using a pick-and-place manipulator for the handling.

Today greater emphasis is being placed on keeping costs down, profits up and staying ahead of competition. To reach these goals it is important to complete rapid changes method and objects.

This leads to a need for flexible manufacturing systems designed for small batches. The tools for this are already available and they include:

- ☐ Advanced industrial robots and NC tools
- ☐ Flexible grippers
- ☐ Computer link
- ☐ Automatic gauging with feed back
- ☐ Automatic tool changing
- ☐ Sensor technology for simplifying peripheral equipment

Fig. 3. A cell in operation.

Putting robots in cells helps the work to flow

Brian Harris, technical sales manager at George Kuikka, explores the merits of using robots to load machines

WITH the introduction of Electrolux (now ASEA) MHU industrial robots into the UK in 1975 economic justification was the main premium for acceptance.

The MHU range of robots were designed specifically for materials handling in the low to medium duty range – up to 15 kg – and therefore a prospective customer did not have the humanitarian or environmental reasons for their justification. Labour was relatively low cost and therefore companies who were interested in robots for materials handling or machine loading applications initially applied the heavy duty or process robots.

'Parts feeding should be upgraded to robot status'

Fig. 1 indicates the increase in labour rates relative to the small increase in the cost of robots which shows that machine loading is now more readily justified.

Several methods of justification are used but a simple rule of thumb seems to apply, in most cases.

- □ 1 shift 1 machine not justified;
- □ 2 shifts 1 machine possible justification;
- □ 3 shifts 1 machine definitely justified.

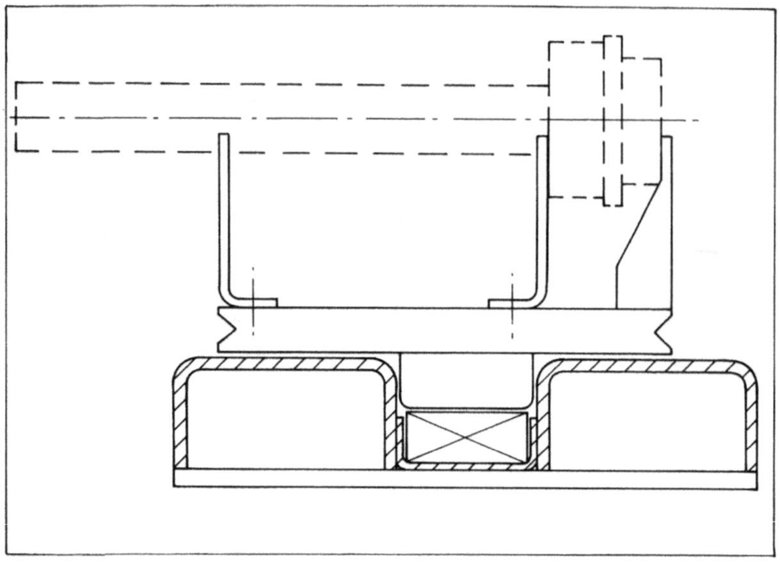

Fig. 2. A typical pallet.

Any extra machines in the cell increase the justification.

Simple payback formula

$$P = \frac{I}{L-E}$$

where

P = Payback period in years

I = Total Capital Investment

L = Annual Labour Savings

E = Annual Expense of Running Robot

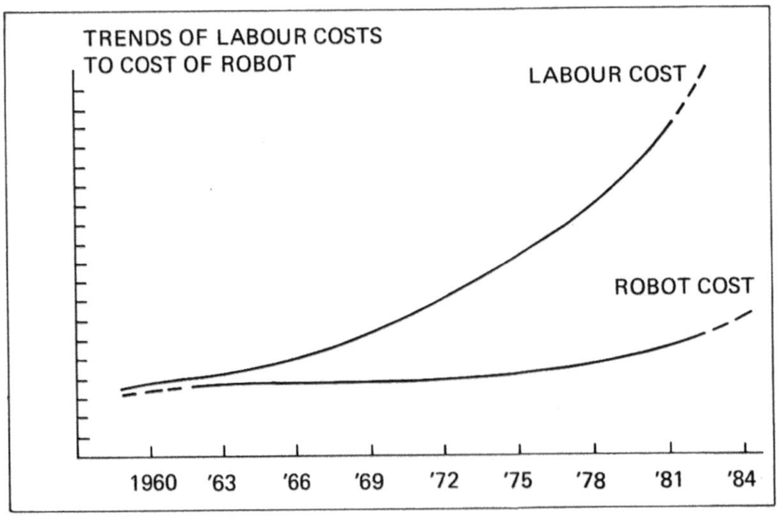

Fig. 1. The trends of labour costs to the cost of robot.

More comprehensive systems are used on a return on investment but it is not necessary to introduce these systems in the early stages. One question often raised when considering return on investment is the expected life of the MHU, which we have proved to be in the region of 10 years. One application in the U.K. has been in operation $3\frac{1}{2}$ years on a 3–shift system ($5\frac{1}{2}$ days/week), which gives an equivalent life of $10\frac{1}{2}$ years.

The benefits of an industrial robot are well known but worth repeating with some additional comments made from users.

- □ Increased Production
- □ Constant Production
- □ Constant Product
- □ Improved Product
- □ Reduction in Unit Cost.

Users have also found benefits that were not included in the original justification, such as:

- □ Increased Tool Life – up to 40% due to the constant production time and allowable increase in coolant flow.
- □ Elimination or reduction of work in progress where machine cells are linked with robots or conveyors.
- □ Flexibility to be re-programmed or re-

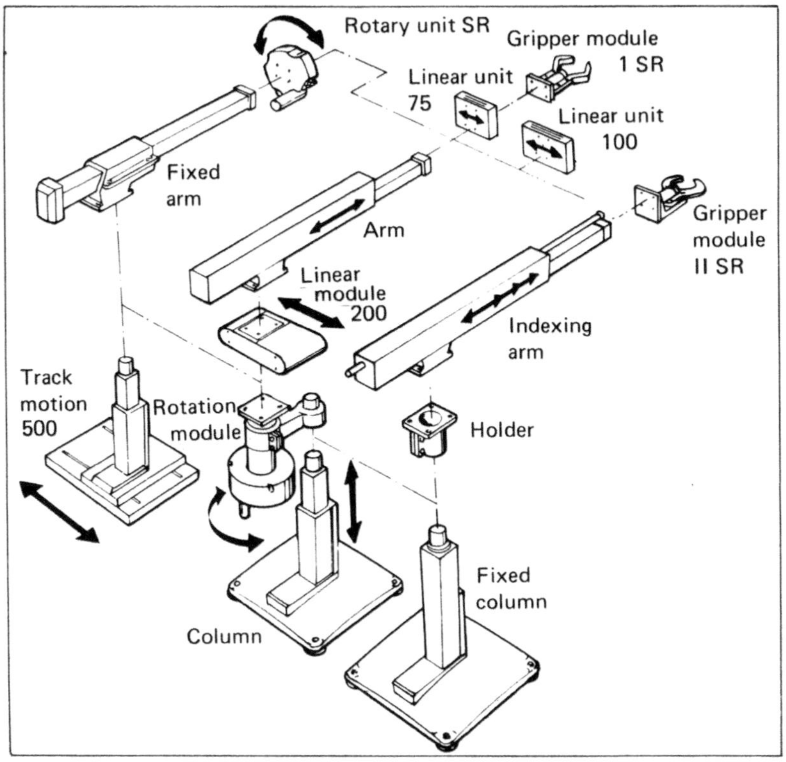

Fig. 3. Configuration of an MHU Senior robot.

tooled to a new product.

Today's industrial robots are not limited by product operation or industry. The MHU range or robots are used in the following industries:–

- ☐ Injection Moulding
- ☐ Die Casting
- ☐ Press Loading
- ☐ Glass Handling
- ☐ Machine Loading, including: Broaching, Deburring, Grinding, Turning, Machining Centres, In-line Transfer Machines.

There are common elements in all these applications which have to be supplied in all cases of machine loading. Unlike injection moulding of diecasting applications components have to be fed accurately to a robot pick up position.

One recent article was entitled 'The Robot is the Easy Part' and this is often the case. The parts feeding is very often considered a less important accessory to the robot but should be upgraded to the same status as the robot and machine and not be considered an accessory.

Failure in the magazine or infeed/outfeed conveyors contribute more to down time than the robots and machines. The machines and robots being standard products built to a high standard of reliability with the magazine and parts feeding devices a one-off special.

At George Kuikka Limited this problem is solved by the use of the EWAB Conveyor System which can be used to link machines automatically with or without robots depending on the complexity of the shape of components – i.e. the multi-conveyor or pallet conveyor.

'High standards of reliability'

Over recent years EWAB in Sweden and George Kuikka in the UK have considerable experience in the application of both types of conveyors on robot applications, but due to the increased complexity of shape of components the pallet conveyor is becoming the standard to be applied to all robot application.

The EWAB pallet conveyor is manufactured to the high standard of reliability required in a robot machine cell and is no longer a one-off special. All component parts, automation and control equipment has been standardised and it is only the pallet that has to be designed to suit the component to be transported. This aligns with the robot having only the gripper to be designed and with the machine tool having only the tooling and fixture to be designed to suit.

☐ The George Kuikka system therefore offers:–

☐ A standard modular MHU robot system supplying only the modules required for a particular application and only the gripper designed to suit the component.

Fig. 4. One of six cells in the robot line at Rolls-Royce.

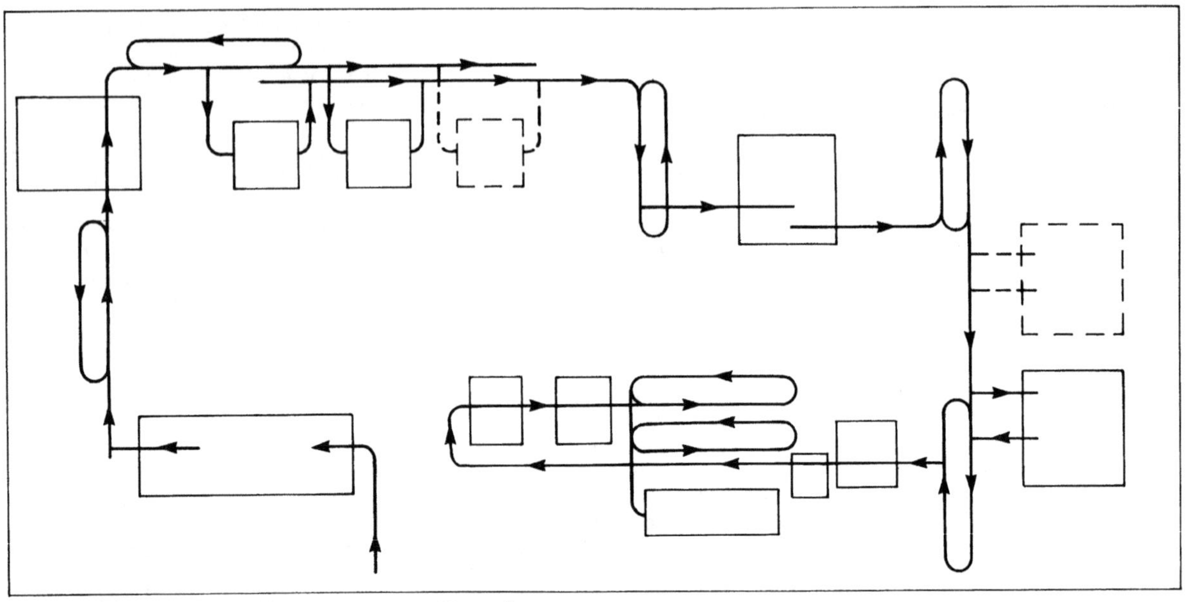

Fig. 5. Conveyor system for a system of robots/machine tools.

- ☐ A standard flexible pallet conveyor System giving transportation, accumulation and accurate location of parts for robot pick-up, with only the pallets designed to suit the component. Lay-out of the conveyor is totally flexible with the ability to turn 90° or 180° bends, to have long horizontal runs and to climb or descend to suit different heights of machines. If required switching and selection etc. are possible with main line computer control.

- ☐ A standard Control System giving sequential monitored control of all robot movements and interfacing with all machines, gauging stations, conveyors, etc. All interface inputs and outputs to machines etc. are connected through a master connecting box into individual connecting boxes on the machines to ensure ease of installation.

By using open contacts all individual machine circuitry is kept clean with no interference from the robot control system. The control system supplied has self-diagnosis, multiprogramme capacity of 127 programmes; up to 250 inputs and outputs, jump facility used when accepting and rejecting components.

All the above standard features are to reduce design, development and installation time by using proven standard modules with total flexibility in design. At times the conveyor is used without robots to link machinery, but this can only be applied where the components can be pushed off the conveyor down gravity shutes into the machine tools as illustrated in Fig. 5.

'The chain will last 93 years'

This system has been installed for at least seven years at Volvo in Sweden and their own figures have indicated that the chain will last 93 years. We have to say that the chain is lubricated by machine coolant carried with the component.

By introducing turnover stations on the conveyor this can reduce the number of functions required on the robot or reduce the number of functions required on the gripper. The capability of the conveyor to bend 180° at a radius of 150mm means that it can be positioned to either turn a component or to bring the component to the required position between robot and machine. We illustrate this with two examples of recent installations in the UK.

Example 1 – Pallet Conveyor feeding parts to a robot/machine cell.

The common elements of a robot application are shown here:

- ☐ Infeed buffer store
- ☐ Industrial robot
- ☐ Automatic machine – suitable for robot loading
- ☐ Outfeed buffer store
- ☐ Automatic gauging station – optional
- ☐ Cleaning or wash station used when gauging is incorporated in cell
- ☐ Perimeter guarding

In this case the Senior MHU Robot which has the capacity to rotate 360° picks up the component from the accurate location on the conveyor (accuracy can be $\pm$ 0.03mm if required) and loads the component through operations one and two then washes prior to gauging.

If the component is accepted it is placed back onto the conveyor, if rejected an appropriate signal will stop the robot cycle. Gauging can be on every component or if required only on a selected number.

The conveyor system can be adapted to feed the component square to the robot arm or in line with the robot arm to avoid unnecessary rotation in the gripper. As the component leaves the cell on the outfeed side a turnover

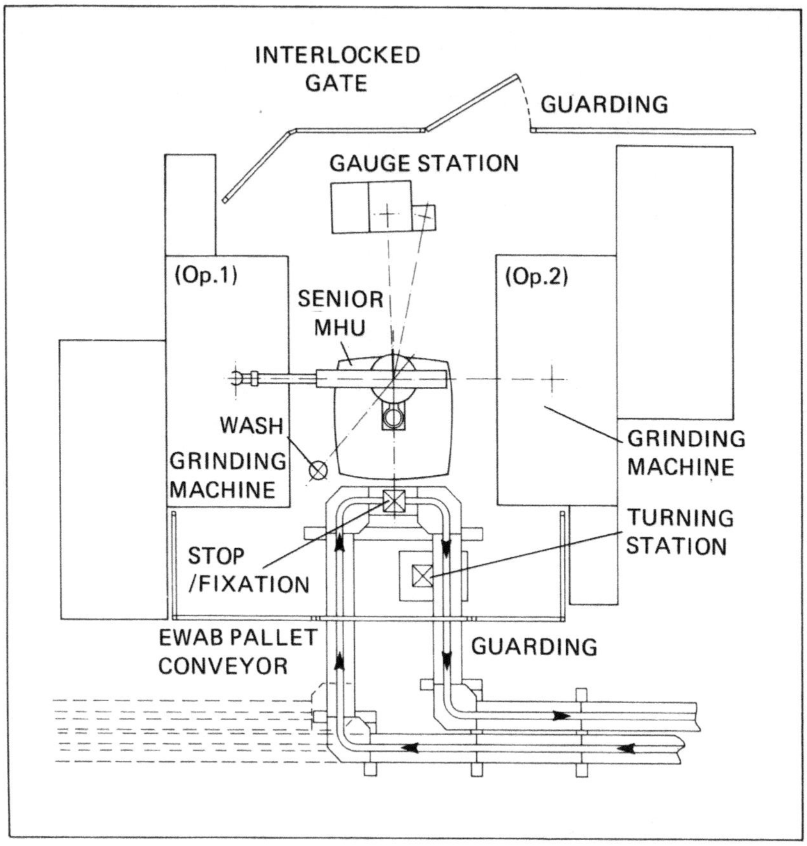

Fig. 6. Single robot cell with EWAB conveyor.

station prepares the component for the next cell, again eliminating unnecessary gripper movements.

The dotted lines on Fig. 6 indicates alternative routing of the conveyor when the infeed comes from another similar cell in line. Each individual robot/machine cell can operate in its own time cycle as a buffer store can be supplied in between each cell.

Thus any down time on any one cell does not stop the complete line. Therefore a buffer store between a number of robot cells in a line is an essential ingredient which the standard EWAB pallet conveyor can provide.

The perimeter guarding is of open mesh 6ft high type with interlocking gate which is the normal accepted

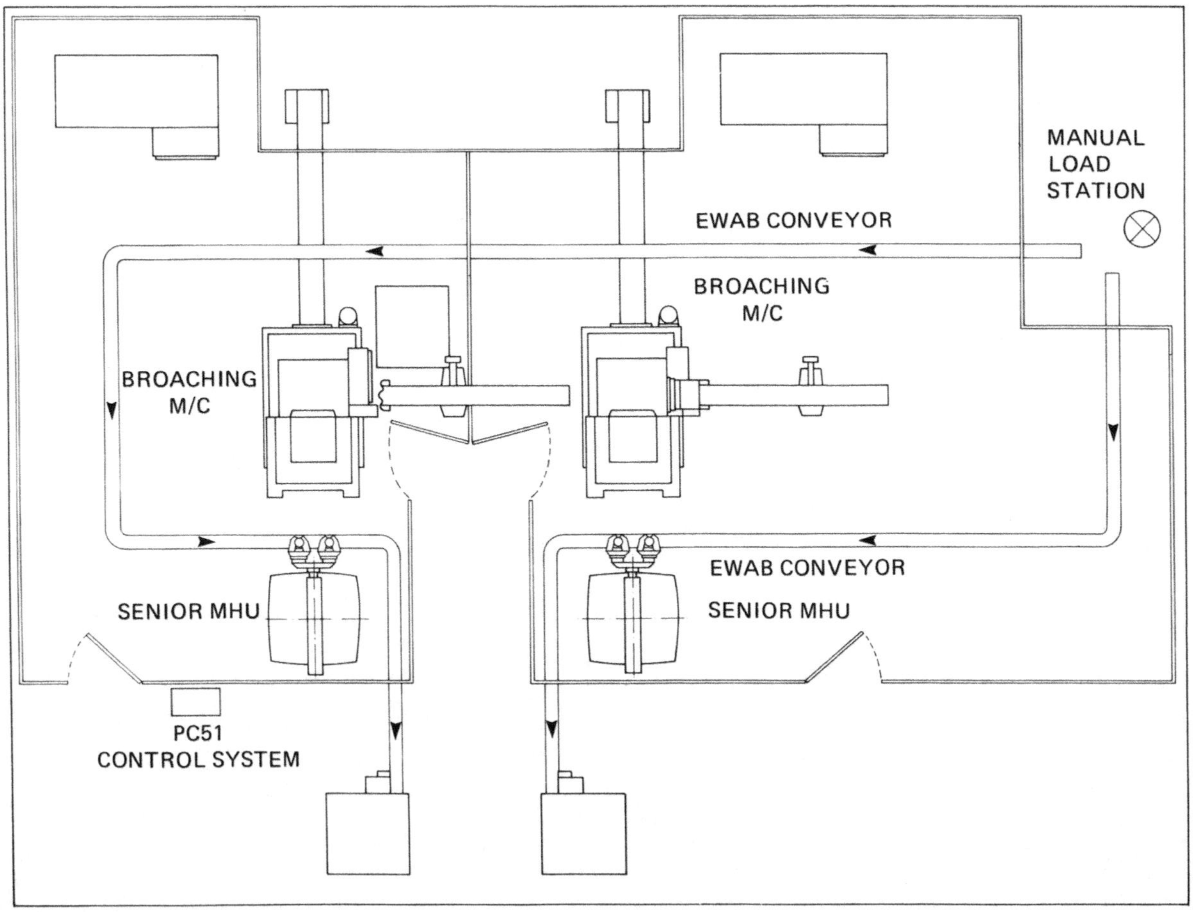

Fig. 7. Twin robot cell with manual load.

Making image processing an attractive proposition for scientists and industrialists alike

Standard software on this user-friendly, desk top instrument provides simple keypad control for:
image storage; image interpretation; image manipulation; image measurement; and a range of image processing functions.

The INTELLECT 100 Image System

- Integral noise reduction (built-in recursive video processor).
- Accepts inputs from VTRs, low light T.V., broadcast T.V., electron microscopes, x-ray devices and others.
- Slow scan inputs available.
- Expandable computer power.
- FORTRAN programmable version available.
- Optional bulk electronic picture storage.

For more details of our range of image systems write for our short form brochure today.

MICRO CONSULTANTS LTD., Kenley House, Kenley Lane, Kenley, Surrey CR2 5YR, England Tel: 01-668 4151 Telex: 946643 MICRO G

μC A member of the Micro Consultants Group

MICRO CONSULTANTS

practice on the majority of robot installations. The interlocking gate has a signature key which isolates the robot and only allows the gate to be opened after a time delay to allow the robot to complete its function. Alternative systems 'park' the signature key in the control system of the machine which means that the removal of the signature key isolates the machine prior to opening the gate.

Example 2 – Multi-conveyor feeding parts to a Robot Machine Cell comprising:– Two robots with twin grippers; Two broaching machines; Two conveyor stations.

This example illustrates the use of the conveyor system to bring the components in front of the machine which eliminates the requirement for rotation of the robot. The only modules required for this application being telescopic column, telescopic arm and twin gripper to feed the twin vertical broaching machines. For this reason one control system had sufficient inputs and outputs to control the two simple MHU robots and the conveyor system. Thus the system was designed to suit a budget limitation.

'Many want to automate'

Designing to a budget is possible with this modular system as the three main ingredients of a robot application are: Standard modular robot; standard modular conveyor system; standard modular control system.

There are many potential users looking to automate their machine loading applications who are hesitating, due to the problems of justification and parts feeding, but we believe the ASEA MHU and EWAB pallet conveyor system answers many of these questions.

Where applications do not justify the use of a higher price high technology robots, the pneumatically operated programmable dead stop robot with its modular concept can easily fulfil all the requirements.

This has been proved on many applications, over 50 in the UK and two users comments can sum up this claim. One high technology robot user after seeing the MHU robot film stated: 'I've bought the wrong robot your MHU could have loaded my machine at half the cost'.

(This user is now looking for a more suitable application for his high technology continuous path robot) and another user stated he attended many robot seminars showing high technology robots when all he wanted to do was to 'simply load a machine'. He approached George Kuikka and its MHU has been loading four components every 15 seconds into a six station Transfer machine for the past three years.

Washing away those blues with flexible manufacturing

Chris Boughton, sales director of CERA International Holdings, looks at robots for washing

FOLLOWING the general acceptance of the use of robots in industry, CERA was determined to research the use of robots into the cleaning systems that it produces. As a company already involved in supplying automation and test equipment, the step into robotics was not a devastating leap.

In the Autumn of 1980, staff began work in developing a Flexible Process System to be compatible with the Flexible Machining Systems at that time becoming available. The basic philosophy in this development was that: Flexible Machining Systems + Flexible Process Systems= Flexible Manufacturing Systems.

The work involved the two main areas of robotic washing, that is,

- □ the robot taking the component to the wash stations, as shown in Figs. 1 and 2. This is used for high production rates.
- □ the robot taking the wash jet (or system) to the component, as shown in Fig. 3.

After an 'Open House' held at the Mitcham Plant, the basic parameters of the robotic cell, shown in Fig. 2, were tested using an IRb 60 ASEA robot. From this work the detailed designs were made for the system.

CERA is planning to demonstrate a similar system in operation during an 'Open House' this year.

Fig. 1 is a typical layout to demonstrate the Cell Type Flexible Process System (FPS). This particular illustration is of a cylinder head but the cell can process a variety of components through a number of modular processes as the process is fully integrated as a programmable cell. The robot controller contains programmes for each component to be processed.

The cell has at its centre the robot. Within the operating range of the robot, modules are positioned in which specific processes are carried out. Other processes can be accommodated within the cell by the addition of further modules. A cell with five modules is described later and illustrated in Fig. 2.

Flexibility is achieved by having a series of programmes available. Selection of the programme appropriate to the component is carried out by either signals from the FMS preceding the cell or by using sensors at the load station.

Flexibility within the module is achieved by programming the tooling to optimise conditions for the process. The robot is used in each case to manipulate the component into pre-determined positions, finally passing the component to the automation thus completing the FMS.

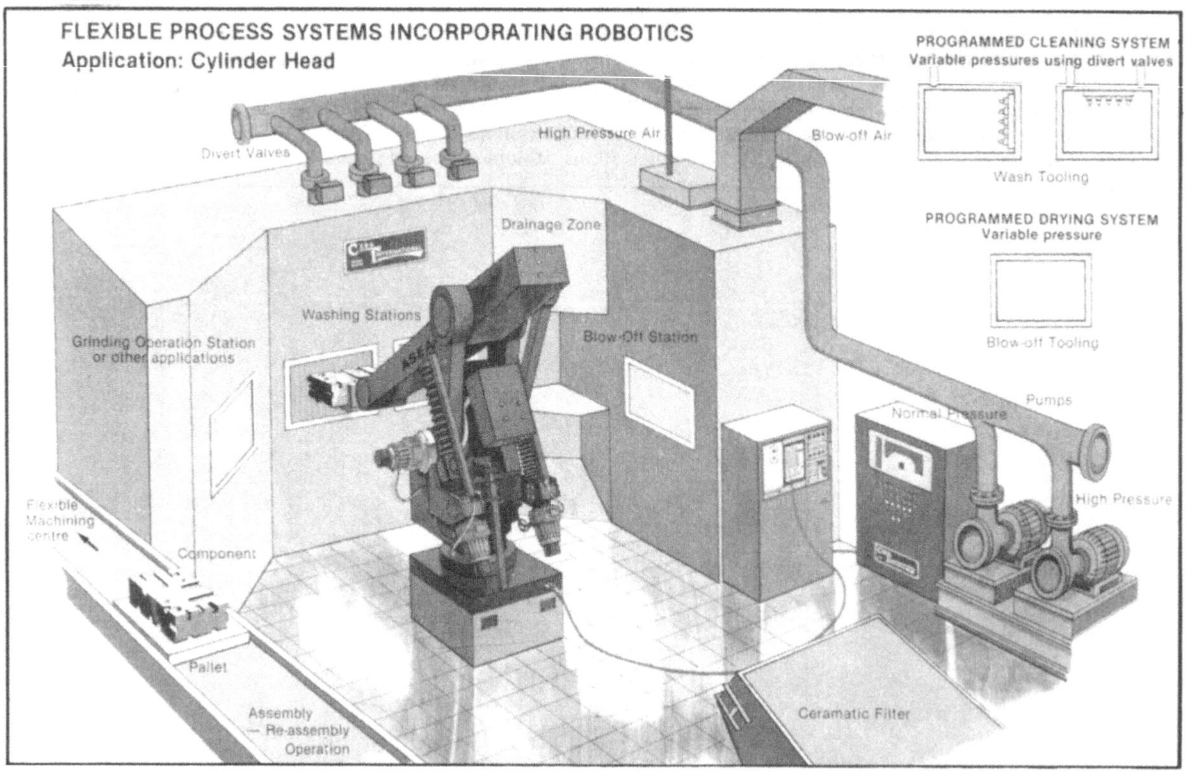

Fig. 1. A typical process systems cell using a robot.

Based on the cell arrangement, the system shown in Fig. 2 has been designed for integration into a line producing 30 cylinder heads per hour. However, in addition to the washing drainage and drying processes the cell in this case carries out four other

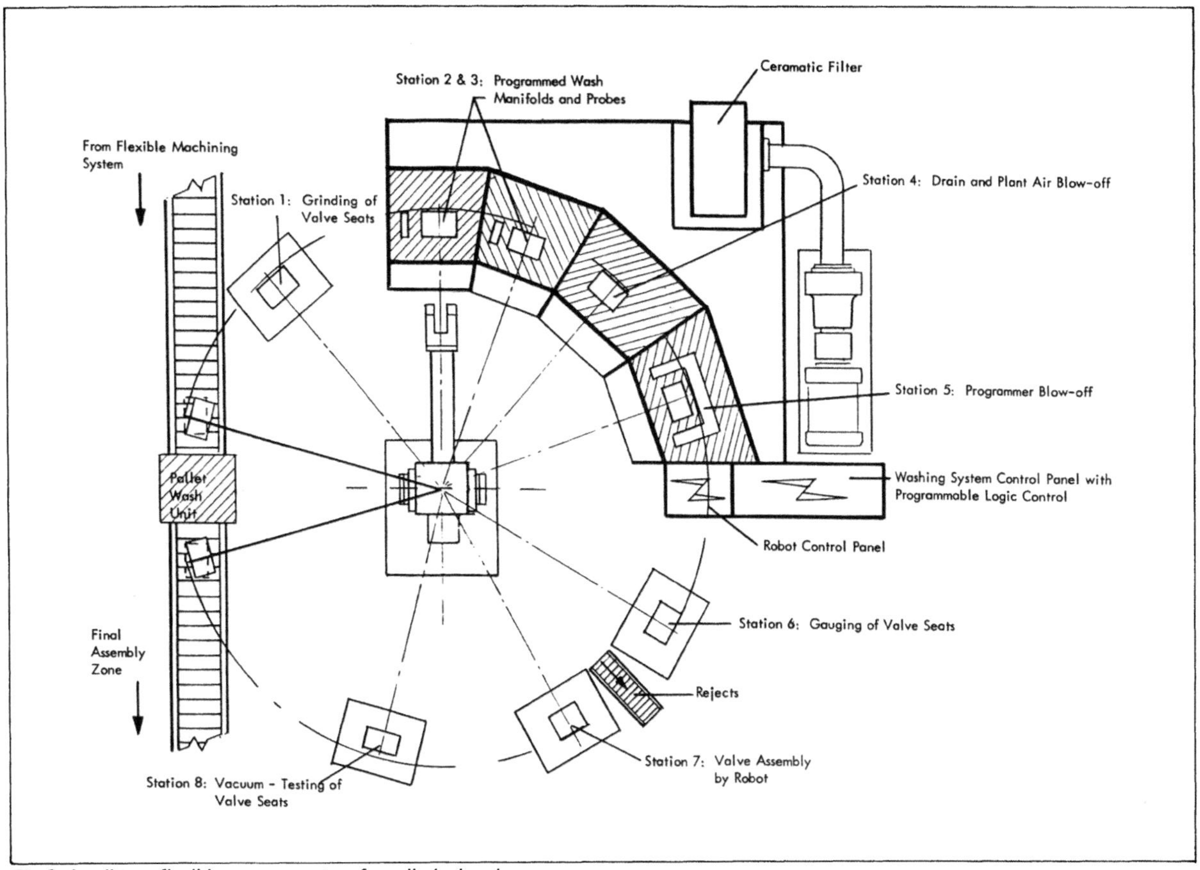

Fig. 2. A cell-type flexible process system for cylinder heads.

processes.

- *Valve Seat Grinding* The valve seat grinding module is equipped with a hydraulically operated variable speed grinding head and an automatic indexing table. The four valve seats are ground independently using a static valve stem arbor. Automatic grinding wheel dressing together with compensation for grinding wheel wear is included.
- *Valve Seat Gauging* Four independent sensors are used in the component gauging module to check the depth to which dummy valves penetrate into the valve seats. If the required penetration is not registered the component is diverted onto a reject conveyor. Each gauge has a digital read-out which automatically shows an 'O' datum when the station is at rest.
- *Valve Selection, Automation and Assembly* The valves are loaded manually into two magazines each with one hour's production time. The valves are selected from the magazines by indexing chain automation and are moved forward for processing. As the inlet and exhaust valves are manufactured from different materials, proximity switches are used to ensure that the correct valves have been fitted to the appropriate magazine. If an incorrect valve is found it is rejected. In the last position of the chain system, the valve is both lifted and accurately positioned for the robot to collect and assemble into the valve stem.
- *Vacuum Testing* In this module an overhead sealing ring is hydraulically lowered to seal around the valve heads, at the same time the valve spring load is simulated on each valve prior to vacuumising the chamber. If a leak is found in excess of the requirements, the component is rejected and is passed by the robot to the reject station.

Despite the fact that there are eight stations in the complete system, the two minute cycle time has been met – with 11 seconds to spare, the robot being operational for 109 seconds per component cycle.

Whilst the robot loads and unloads each of the stations in the cell to complete the processing of the component the pallet is passed through a cleaning and drying station on the automation line to prevent recontamination when the component is reloaded.

When developed this particular application was the most demanding

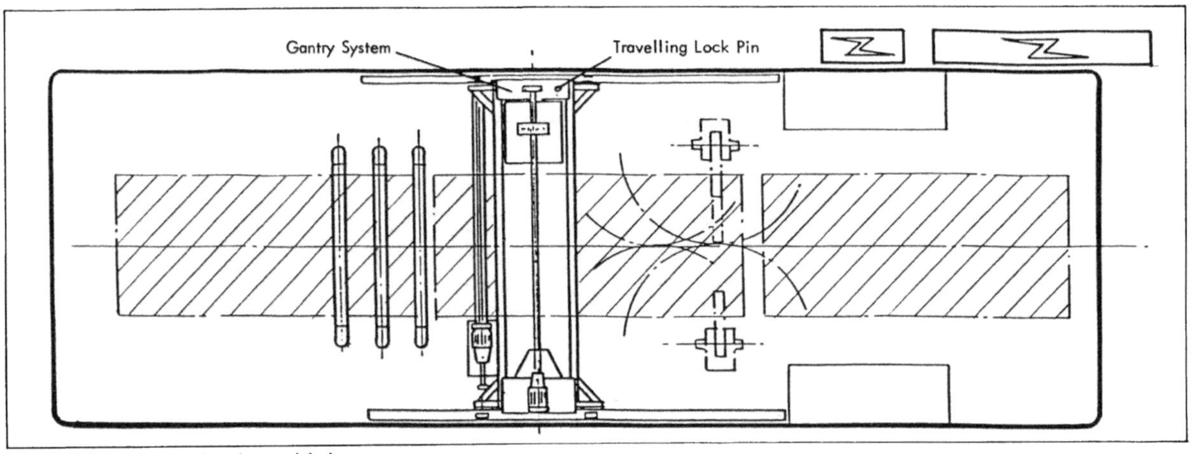

Fig. 3. A through-type cleaning and drying system.

that our applications engineers had undertaken requiring as it did the full complement of application technology, that is:
Floor-to-floor (cycle time) economy.
Interfacing.
Load/unload interfaces with machine tools.
Manipulating programmes for drainage.
Palletisation of the components.
Plant Layout philosophy.
Process interlocks.
Specially manifolded cleaning and drying.

The alternative arrangement to that shown in Figs. 1 and 2 is the use of the robot to directly spray wash liquid and drying air onto the components. Fig. 3 shows a robotically assisted cleaning and drying system for assembled engines.

In the robot assisted through-type engine assembly cleaning and drying cell (Fig. 3) the engine assemblies – there are about 20 variants of V12 and V16 models in the range – complete with radiators, breathers, and so on, enter the machine on fixturised trolleys. These pass a component sensor which detects which of the 20 variants is entering the machine and instructs the micro-computer based controller and robot controller to use the appropriate programme.

In the robot section (centre) the component is presented at a precise location with respect to the robot which in this instance is mounted in the inverted position on a traversing gantry to enable it to move to the required position.

According to the instructions received from the sensor the robot selects one wash jet from a series of specifically designed jets located in a special fixture. With its three-dimensional dexterity it can move the jet in and around the complex-shaped component to remove the soils remaining after conventional cleaning. The robot replaces the wash probe and moves to its hold position after completion of the wash cycle.

The component is moved back to the first third of the machine in which three rings of stainless steel jets pass water and iron phosphate over the component whilst it is reciprocated backwards and forwards.

After completion of the washing and phosphating processes the component moves forward to the blow-off section located in the last third of the housing. On its passage to the blow-off section, the component passes a ring of jets which rinse off the excess phosphate solution.

Fig. 4. An overhead-mounted robot for washing.

The engine proceeds to the drying section in which it is reciprocated within rings of air drying jets before returning to the robotic section. Its arrival in the robot section initiates the robot to select an air jet probe which it uses to blast any remaining water from the cavities which the conventional air dryer was unable to penetrate.

'The use of robots in industry is commonplace'

The protection of the robot within the potentially hazardous environment of an industrial washing machine is an important factor in robot selection. This is an area of robot design that some manufacturers would do well to look at further.

As new areas of robotic use are found in industry, the robots employed will be required to operate in the most arduous environments.

Selection of a robot is an area in which both application engineers like ourselves and end users, need as much comparative information as possible. CERA is happy to see the progress that the British Robot Association is making in this field. The Members Handbook 1982/83 is a useful document in industry's endeavour to select the correct robot for a particular application. It is still difficult however for end users to keep pace with the rapid developments in robot technology, in particular in the field of micro-processor technology and controls.

The use of robotics in industry is now commonplace and the benefits of rapid information with regard to production, stock control, monitoring and diagnosis are already being accrued. True benefits of flexibility in production are a more complicated matter.

FMS together with FPS offer the greatest rewards to industry, but like most things in life they require the greatest investment. With thought to component design and common handling by pallets, for example, a variety of components may be processed. At this time, and with the application engineers who thoroughly understand the processes involved, the real step forward in Flexible Manufacturing will be achieved.

No novice in surface coating

Arvid Hareland, of the robot division, Trallfa, reports on the state of the art

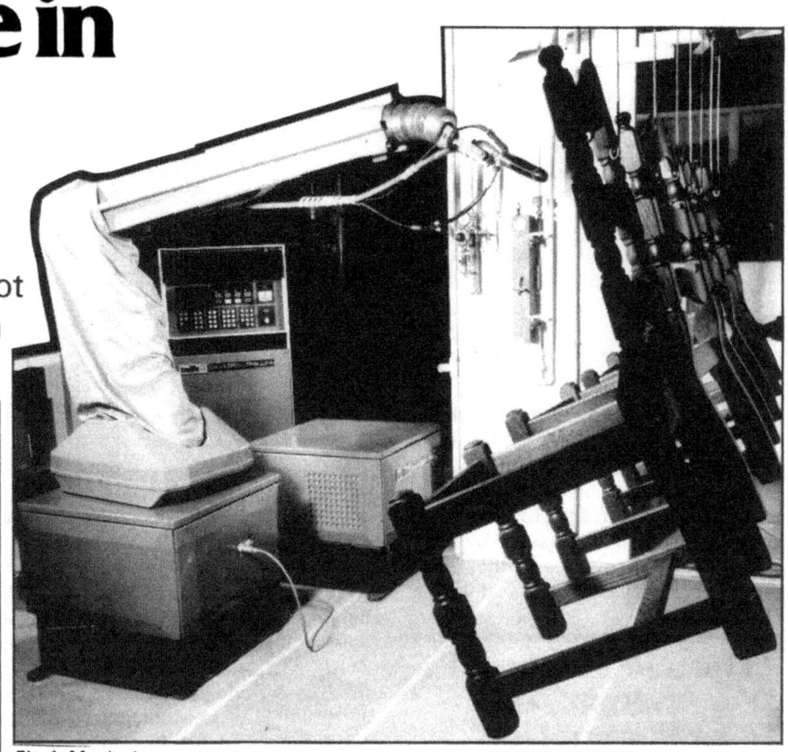

Fig. 1. Manipulator coats chairs.

INDUSTRIAL robots continue to find new areas of application in welding, handling, casting, parts assembly, surface treatment, surface coating and many other areas. With the present development of controller units, improvement of dynamic characteristics, and other advances, robots are able to perform more jobs quicker and better than the human hand. Surface coating is one area where the robot is no novice, even if it is now becoming able to perform more demanding tasks even here.

What is surface coating from the robotic point of view?

Surface coating generally means applying a substance to a piece of work and covering it. Seen from the point of view of robotics, however, surface coating is something more. The purpose is still to apply a substance to a piece of work, but we must take into account the fact that the robot is an automatic tool which must function together with the rest of the production. For this reason, we must be more technical in our description of surface treatment.

'Robots must know when objects enter the field'

The object to be coated must be brought into the robot's field of operation before spraying and out of the field of operation afterwards, so we must have a transportation system. Further, the robot must know when the object has entered its field of operation and when spraying may start. This requires a start-up signal from the transportation system.

When the start-up signal has been given, the substance is applied. The robot must then be able to move the spray gun in specific trajectories and to control the on/off switches in such a way that the surfaces of many different types of parts, from simple plates to parts with complicated geometrical forms, may be covered properly. It must be possible to program these different movement patterns into the robot.

If several colours are used, it will also be necessary to have a colour switching system which jets out old paint, cleans, and fills the supply pipe to the spray gun with the new colour. If several types of products and/or several colours are sprayed in the same production line, the robot must be fed with information about which program to run and which colour to use.

Which abilities are necessary for the robot to live up to expectations when used for surface coating?

The first and obvious point is that it must be able to carry the heaviest automatic spray guns at the extreme end of the arm without this leading to inaccurate positioning, even with maximum acceleration.

For robots where programming takes place by the 'lead-through teach-in' method, as is the case with the Trallfa robot, it is also necessary for the robot arm, with gun mounted, to be easily led through the different movement patterns which are needed to achieve a good surface coating of the different parts.

Some companies use robots for spraying a large number of different parts, and these need considerable program storage capacity. This will also make the robot more useful in the sense that companies producing short series of many different products may have an opportunity to use robots.

If the robot is to be flexible with regard to work tasks within 'surface coating', it is also a requirement that the programming can be done both through a continuous path and in point-to-point mode.

It is necessary also for the robot to have a wide reach so it can cover large items. This must be attained even though it may frequently be necessary to place the robot in an existing spraying cabinet with limited dimensions. In order to spray items with complicated geometry, it is essential to have great flexibility in the movements of the robot, hence a large number of degrees of freedom.

To make a good surface coating, a high degree of accuracy is required for some types of spraying. As in many instances the robot will be working towards a conveyor in constant motion, and therefore has severely limited time for spraying each item, the combination of high accuracy and high speed are required.

The fact that the robot is working with a conveyor whose speed in many cases is determined by other parts of the production process means that the robot must be able to adapt to the conveyor speeds in use. The conveyor speed will also usually vary with time so it must be possible to synchronise the robot with the conveyor.

When the robot is to operate within the total system in a company, it is also a requirement that the robot must be able to communicate with its surroundings, and it must be possible to integrate it into an overall control system.

The day-to-day operation of the robot is usually undertaken by an operator who is not a robot specialist. He may perhaps be a finisher. The consequence is that the robot must be adaptable to the operator, and it should be equipped with self-diagnostic facilities to the greatest possible extent.

If errors should develop during spraying, it is also important that the robot

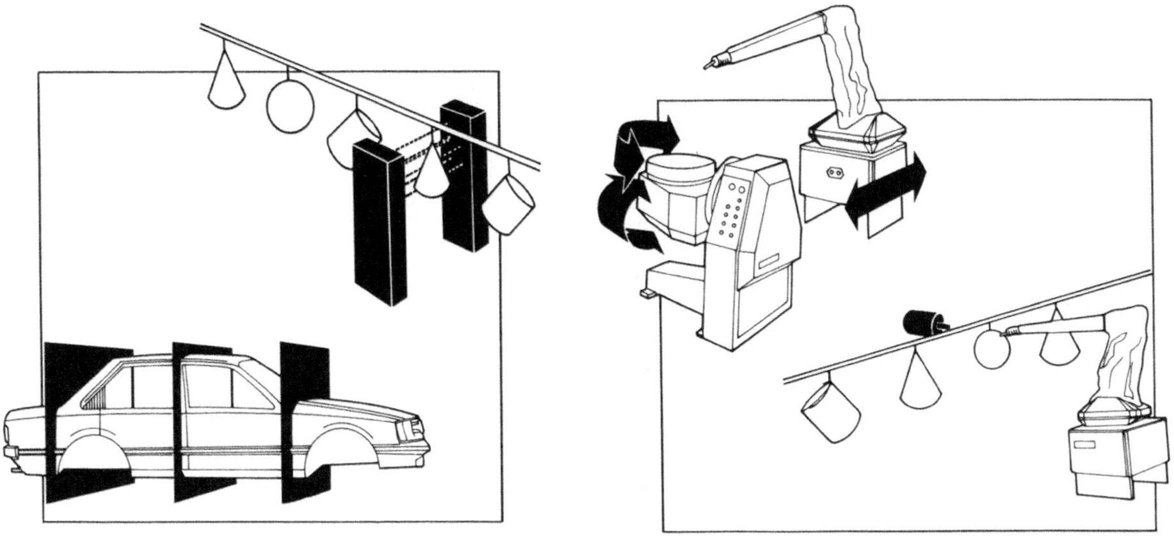

Fig. 2. Various techniques for painting components.

behaves correctly. It must be evaluated on the basis of the proper course of action in each situation. From this point of departure, the robot must be given a defined, correct behaviour if the power supply should fail, if the controller should fail, in case of collision between the robot and the item, and so on.

The robot, on its part, must be able to accept the physical environment to which it will be exposed. It must tolerate different types of noise, paint and similar particles in the air and, not least, strong solvents which may easily affect hoses and cable insulation.

Which, then, are the main problems encountered when using robots for surface coating, and how have these been solved?

One of the major problems is undoubtedly the spraying of surfaces with very complex geometrical forms. If surface coating with a robot is to produce a fine, even surface all over, very complex movements of the spray gun at the extremity of the horizontal arm will be required. It is difficult to program this through equations and ordinary machine instructions.

The best solution is then that an experienced spray finisher will undertake to train the robot in the movements which he knows will result in a good surface. This method is called 'lead-through teach-in'. Dare we say that surface coating with spraying may be half science, half art, and that the art part of the job is preserved in the automation when an experienced operator trains the robot in this way.

When the robot is to be taught in this way, it is important that the robot is easy to move through the training phase. This has been accomplished by the Trallfa robot.

As already mentioned, the robot will be placed in a rather particular environment when used for surface coating. Electrical systems may easily cause sparks which may ignite the explosive gases found in spraying cabinets. This danger has been eliminated by the hydraulic system in which the electric pump motors are physically isolated from the manuipulator, and which may therefore be put in a safe place. Furthermore, the electrical wiring of the manipulator has been made intrinsically safe by the use of zener barriers and minimal current requirements.

If the item to be sprayed is large, it may be advantageous to divide the spraying program into several modules. If the conveyor stops, the robot may finish spraying the module currently in progress and then terminate.

The possibility of modular program development has also alleviated another problem. Take-up of spraying programs for large parts could be very time-consuming as long as it was necessary to retake the whole sequence if errors were made near the end. Dividing the program into modules means that only one module needs retaking if errors are made.

In many companies, the robot is used to spray a number of different parts with a number of different colours. The different program numbers may then be queued for the robot, either manually or automatically. If a colour is to be replaced, the robot may control the valves for the colour exchange system and so on, in programmed sequences.

Which surface coating applications are now possible for robots?

Between 50% and 60% of all Trallfa robots used in this application are supplied to the motor industry, including component suppliers. They use the robot for finishing both parts and whole car bodies, and both base coating and top coating are sprayed on. In addition, the robot is much used for spraying PVC coating on car chassis.

Another large user of robots for surface coating is the sanitary industry, which has close to 20% of all Trallfa installations within surface coating. These robots are used mainly for enamelling or glazing bath tubs, wash basins, WCs, and so on.

The third large user group is the domestic appliance industry. The robots are used to finish products such as fridges, freezers, washing machines, etc. This industry has about 10% of Trallfa's installations for surface coating.

The robot is otherwise used for finishing a great many types of products. It is used for spraying lubricants, spraying insulation materials and a number of other tasks, large and small. The number of areas of application is, it appears, enormous.

Putting on the style

Dr Tony Cronshaw, senior engineer at Patscentre, explains the use of robots in the food industry

Piping chocolates by robot.

THE food industry is a major employer of labour, often in repetitive jobs, and is fast realising that robots can help. For example, Gerhard Schubert's 'Roby' robot shown at the Interpak 81 exhibition in Dusseldorf provoked much interest picking and placing chocolates at rates up to 90 pieces a minute. And it is the flexibility of programmable robots to cope with batch changes in the packing area that looks set to make the first impact.

New developments in 'intelligent' robots are also promising to automate many jobs where until now the human worker has had the edge. Using machine vision, newer robot systems can combine basic inspection tasks with picking and placing.

Of the four basic areas of food processing – raw material handling, processing, packing and warehousing – the last two are the main areas for robotic applications in the sense of robots as we generally know them. Automation in a wider sense is steadily being introduced however across all four fields and progressively this is becoming more flexible.

Automation in the warehouse is now a major growth area. After a shaky start in the late 1960s, there has been a resurgance of interest in automated warehousing systems. Equipment, often unreliable in the early days, has been transformed thanks to cheaper and more reliable control electronics.

Also, the advent of data coding, including the 'best before date', in the EEC since January 1983 together with bar coding now means that packages and pallets are now inherently labelled externally. This information is often what the automatic warehouse system needs for product identification and computer storage. The overall result is high density high efficency warehousing with minimal labour.

Palletising using robots is a particularly attractive application well suited to robots. Where flexibility is important, for example, for batch changeover from one pallet pattern to another a robot can prove highly effective.

The Robopal system by Mepal is a pick-and-place robot specifically designed for palletising. Handling 300 packages per hour it is also easily reprogrammed for product changes through a cassette storage system and microcomputer control. And the Webster Griffin robot palletising systems, though not robotic in the usual sense, are nevertheless flexible and programmable and can handle up to 2,500 packages an hour.

Machine vision is set to make a big impact in raw materials inspection. Machines for cereal, meat and vegetable inspection are already available. The Glafascan system computes the fat/lean ratio of raw meat enabling the fat content of products to be more carefully controlled. And Gunson-Sortex offer systems for the sorting of cereal grains, rejection of foreign bodies and so on.

Intelligent robots using machine vision for guidance are now becoming available and early prototype trials are already underway in the confectionery industry. Whereas a basic robot operating with a fixed pick up point is unable to cope with slightly mis-shapen objects a robot with machine vision is able to adapt automatically. Sensing is accomplished by television cameras or linescan cameras and a microprocessor is used to compute the position and nature of the object in view.

The automatic chocolate decoration system shown at a prototype level by Patscentre International uses this new technique. A linescan camera is used to pick up the position and orientation of chocolates which can be on a moving conveyor or on moving factory trays.

'Benefits of automation are under-estimated'

A robot arm equipped with a decorating nozzle is then commanded to extrude piped chocolate patterns onto the tops of the chocolate. Each decoration is precisely aligned even though chocolates can be anywhere in the tray and rotated through any angle. The system can even detect mis-shapes including undersize and oversize units.

An extension of this technique under investigation by Patscentre International is for an automatic chocolate packing system. Again a machine vision system is used but with a higher emphasis in this case on inspection of shape.

A robot arm is guided to pick-and-place chocolates working directly from factory trays. Thanks to the built-in inspection facility, mis-shapes falling outside pre-programmed limits of acceptable shape and size are automatically rejected.

The benefits of increased automation are often under-estimated. In actual fact, the increased efficiency can be quite dramatic. Robots do not need tea breaks, lunch hours or holidays, nor do they go sick or at least not often! Moreover with increased automation and reduced manning, shift working 16 hours or 24 hours a day can become a viable proposition from a manning point of view. Plant efficiency is then transformed literally overnight.

Megassembly: The sleeping giant of robotics

Philippe Villers, president of Automatix Inc., looks at the future assembly capabilities of robots in industry

MOST observers of industrial robots agree that the most promising application of robots is in assembly, an area, presently in its infancy. Assembly robotics is as yet, untested or underdeveloped because of the relative complexity of the tasks and the difficulty in competing with that remarkably agile tool, the human hand, particularly when guided by the human eye and commanded by the extra-ordinary processor that is the human brain. Nonetheless, great strides are being made and will continue to be made as the sleeping giant of robotics –assembly robotics – evolves.

What stands between robotics and its mass use in assembly? In some instances, direct cost presents barriers. In addition, there is growing concern about a more subtle question, one frequently described as 'the question of robustness', that is the ability of the robotic system to tolerate normal variations of parts and all other conditions of assembly.

There are many applications in assembly which robots with ideal tooling and fixturing could, in principle, perform. However, within normal industrial variations and without the human's implicit visual and/or tactile inspection, the robots are not robust enough to continue to work for prolonged periods. Often, to deal with the variety of required assembly parts, they have to be cluttered up with specialised material handling and fixturing which is extremely inflexible, expensive to design, and time consuming to develop.

To overcome these limitations we believe that artificial vision, the 'king' of the senses, complemented by tactile capability and by highly sophisticated computer programs to integrate the sensory input and determine responsive strategies is destined for a major role.

'The birth of megassembly'

We would like to introduce the new term 'Megassembly', to describe robotic assembly processes where the number of assembly stations is significantly large – at least half a dozen –and where the stations must work together to complete a single assembly. Examples of this in the West are still very rare.

However, there are Japanese applications which use simple robots such as the Scara family of robots of which an example is shown in Fig. 1. These are now used together in quantities of 50 or more in such applications as inserting a variety of components, most of them non-standard, on printed circuit boards for manufacturing high-fidelity stereo parts, or inserting electronic and mechanical components such as the bulk of the parts mounted on a video tape deck. This last task is reputed to be completed with an array of over 50 simple robots at another Japanese manufacturer – (JVC).

Robot Megassembly differs from other contemporary robot assembly systems in that most assembly robotic tasks involve one, two, or three robot arms which perform a *portion* of a complete process. Megassembly, on the other hand, involves 10 or more robot arms which may, in progressive steps, complete *a total assembly task*.

The economics of total robotic assembly are particularly attractive; much of the cost associated with robotic assembly and the engineering of robotic assembly applications relates to the feeding of parts and the maintaining of position or orientation, a task-performed by (and hence a cost borne by) the first station of a series. Thereafter, the cost rises slowly as the number of similar operations in the series increases.

Megassembly – through division of labour – allows simpler and faster operations at each station.

Similarly, the computing system, particularly in a network structure, has a substantial cost for the first unit; and thus cost per station is a strong inverse function of the number of stations. In Megassembly operations a centralised station may be used to sort and orient the parts rather than having the full capability at each assembly station.

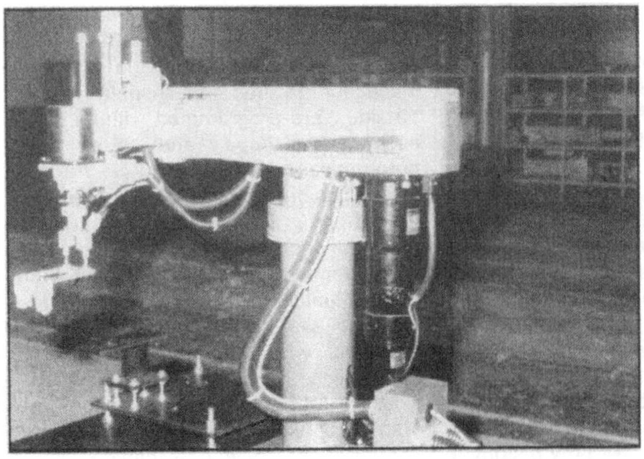

Fig. 1. SCARA robot by Hirata.

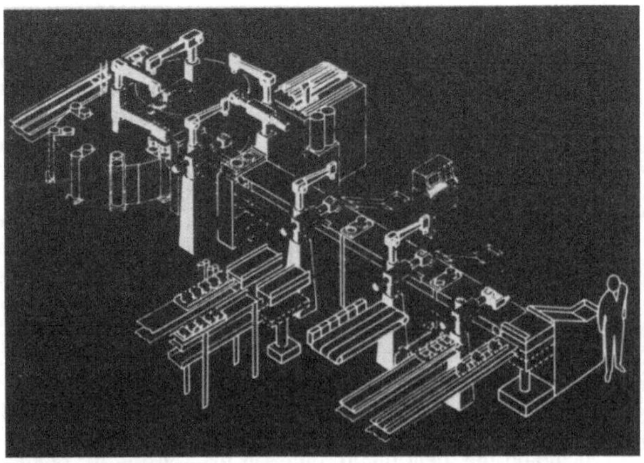

Fig. 2. Westinghouse APAS.

Also, in Megassembly (and to a lesser extent in other robot assembly operations), the classical distinction between assembly operations, and testing, inspection, or other non-assembly processes, is likely to disappear. In an integrated assembly process (the human operator no longer doing the assembly), artificial vision systems will increasingly do the visual inspection and provide real time feedback to the assembly system.

Some processes not normally associated with assembly, such as bonding, welding, and other permanent assembly techniques, or even coating can be conveniently intermixed, particularly in a Megassembly system. Megassembly systems can either be of the progressive line variety or be composed of undifferentiated assembly stations with asynchronous random or non-random (fixed path) part distribution and material handling systems for transferring parts between stations.

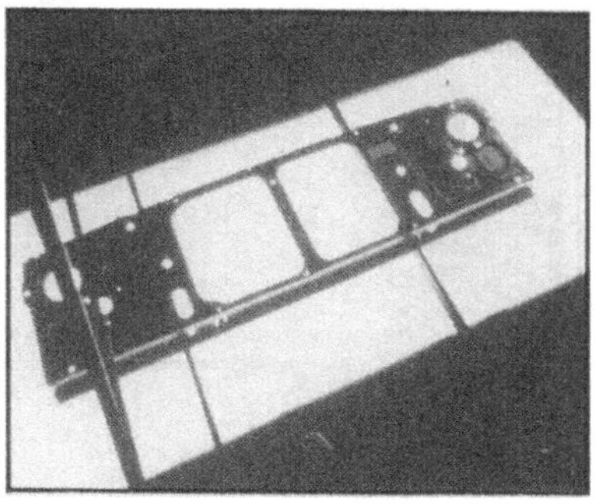

Fig. 3. Inspecting 2,200 holes/min.

Asynchronous material transport systems, regardless of type, are preferred in assembly due to the difficulty of synchronising the robot operation with a moving line – the inaccuracies involved in controlling the dynamics of moving bodies, and to the limitations of working with a fixed time interval rather than a variable time to perform an operation. This is especially important when there is a significant workload variation as discussed below.

One of the economic difficulties associated with assembly is that the continuous assembly of a single model is not economically feasible in many operations. Therefore, robot assembly lines and particularly with Robot Megassembly, need to allow assembly of different but related model parts on the same line without requiring set up changes.

This in effect is, at the limit, an application to assembly of 'Group Technology' thinking where assembly tasks requiring the same assembly station capabilities and processes can be intermixed on the same line as demonstrated on a pilot basis in the Westinghouse APAS system shown in Fig. 2.[1]

In this pilot project, a series of robots and vision modules using the same hardware but with automated tool changing capability and software control assemble a family of electric motors of different sizes.

To be readily adaptable to a range of tasks, robot assembly systems put

```
Enter Password Followed by RETURN
Quick Inspection Startup?
 (Y)es or (N)o
 Type First Letter of Option Followed by RETURN

Default Data File-Name = CHHOLES
Change?
 (Y)es or (N)o
 Type First Letter of Option Followed by RETURN

Begin Training Procedure?
  (Not Training Means Using Existing Data)
 (Y)es or (N)o
 Type First Letter of Option Followed by RETURN
Alter Parameters or Tolerances Before Training?
 (Y)es or (N)o
 Type First Letter of Option Followed by RETURN
->TRAINING

Next Camera is 1
Quit Training Now?
 (Y)es or (N)o
 Type First Letter of Option Followed by RETURN
TRAINING: PASS 1 OF 2
Use CAMERA 1 ?
 (Y)es or (N)o
 Type First Letter of Option Followed by RETURN
Waiting for PART-READY on
Full Field of View for Pass 1 Picture?
Change?
 (Y)es or (N)o

 Type First Letter of Option Followed by RETURN
Present Threshold = 10
Change?
 (Y)es or (N)o
 Type First Letter of Option Followed by RETURN

Waiting for PART-READY on

Area = 408
Store this Hole?
 (Y)es or (N)o
 Type First Letter of Option Followed by RETURN
Use Area-Count Window?
 (Y)es or (N)o
 Type First Letter of Option Followed by RETURN
Quit doing this Option?
 (Y)es or (N)o
 Type First Letter of Option Followed by RETURN

Area = 434
Store this Hole?
 (Y)es or (N)o

 Type First Letter of Option Followed by RETURN
Use Area-Count Window?
 (Y)es or (N)o
 Type First Letter of Option Followed by RETURN
Quit Doing this Option?
 (Y)es or (N)o
 Type First Letter of Option Followed by RETURN

Set up Another Window
 (Y)es or (N)o
 Type First Letter of Option Followed by RETURN
```

Fig. 4. RAIL language for inspection.

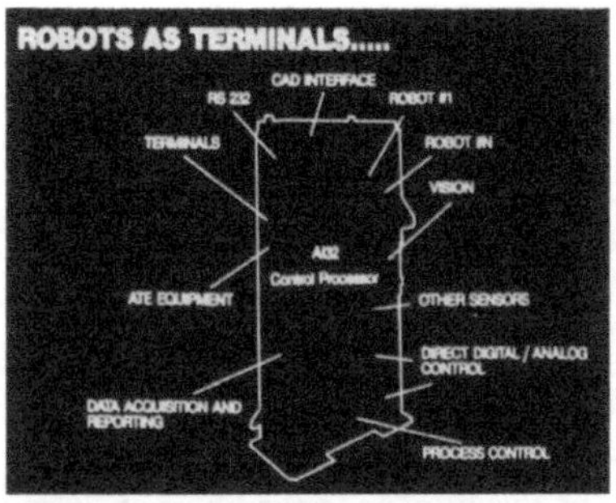

Fig. 5. Functions of A132 controller.

Fig. 6. SI-Handling Mini-Cartrac.

Fig. 7. Volvo guided vehicle.

Fig. 8. A132 robot controller.

significant requirements on robot languages which in turn require the full power of the modern microprocessor to implement. These systems demand large memories and sophisticated higher-level languages to provide the required flexibility.

Advanced robot assembly languages must accept and interpret sensory input, including vision, touch and in some cases force and torque, and yet must be suitable for use by non-programmers on the factory floor.

An example of such a modern language is the Automatix RAIL* language. RAIL is specifically designed to accept vision input and is robot task independent; the language is currently being used in inspection using artificial vision processes, in robot assembly with or without artificial vision, and in such unrelated tasks as arc welding.

The RAIL language is an 'interpretive' language which takes many of its useful features from PASCAL and is intended for the non-programmer. It lends itself to flexible modules and to interactive menu-driven use allowing programs to be created by the operator mode or a CRT screen. Such an example for the inspection task in Fig. 3 is shown in Fig. 4.

Sophisticated assembly requires sophisticated computing. This is particularly true of Megassembly which requires the integration of a number of robots. As a practical matter, this can only be accomplished by some form of hierarchy among the interconnected computing nodes or processors.

With the recent development of modern microprocessor communication and various communication protocols for handling both low- and high-band width with either low- or high-utilisation factor, this problem has been increasingly resolved by the

*RAIL is a registered Trademark of Automatix Inc., Billerica, MA., USA.

computer industry and is ready for adaptation to robotic systems, using controllers which reflect the state-of-the-art in computer sciences.

An example of such a controller is the AI32. The AI32 supports the functions shown in Fig. 5 using up to a million words of direct memory. It utilises the Motorola 68000 microprocessor which has 32-bit addressing and is capable of using higher level languages.

Transporting materials in a multistation or Megassembly robot application, requires a sophisticated material handling system which can either proceed from station to station with 'skip station' capability or accommodate random sequence. Examples of both are emerging. An example of such an asynchronous system is the SI Handling System shown in Fig. 6. An example of the random sequence capability is the Volvo system shown in Fig. 7 – which is an 'automated guided vehicle' with the ability to follow a line on the floor and to be shunted or stopped by radio command.

'Powerful sensory system'

Probably the most powerful and versatile sensory system is the artificial vision system which is now available from some manufacturers as an integral portion of the robot controller, such as the AI324™ These systems support a large number of cameras, the one shown being capable of 16 independent camera inputs with processing speeds of over 10 parts per second for simple geometries and applications. In multi-camera installations the cost per camera becomes correspondingly low and, because of the shared processor, can cost as little as $3,000 to $5,000 per station depending on the type of camera.

Tactile sensing is less developed. In its simplest form, a photocell or a wire which completes electrical contact determines the absence or presence of a part or surface. In its more sophisticated form, 'artificial skin' uses an array of sensing elements, in effect collecting a tactile 'handprint' which identifies the location and/or identity of the parts by its pressure pattern.[2]

Such sophisticated applications of tactile sensing are likely to come into use in the next several years, but will compete with the vision systems. In those applications where vision systems are required anyway, there may be insufficient justification for supplementing them with a complex tactile system. Conversely, where vision is not otherwise required, tactile sensing may prove to be significantly less expensive.

An example of a large size robotic assembly using vision is shown in Figs. 9 and 10. This illustrates an experimental use of an Automatix Cybervision System with vision to tighten adjustment bolts in a car door assembly. A series of bolts for adjustment of the window roll-up mechanisms are mounted in slots and so the exact location of the bolt head cannot be predicted in advance. The vision system is required to guide the robot to each head.

A second automotive application

Fig. 9. Car door – vision aided assembly.

Fig. 10. Close-up of car door assembly task.

Fig. 11. Multi-part feeding.

Fig. 12. Keyboard robot assembly.

involves an engine block on a line where the position of the block cannot be predicted to the required accuracy to complete the mounting of accessories. Vision locates the block so the robot can find its target.

Of course, a discussion of assembly robotics would be incomplete without a discussion of the remote centre compliance[3] which eases the task of inserting tight fitting elements such as a shaft into a bearing. This device originated at MIT's Draper Laboratory in Cambridge, Massachusetts. It is a passive element and therefore not truly a sensor, but is a powerful assembly aid; performing a function which otherwise would require sophisticated sensors. In the case of the IRCC (Instrumented Remote Centre Compliance)[4] sensing and passive self-adjustment are combined.

The question frequently comes up whether it is easier to have the parts pre-sorted and positioned before the assembly station at some central point, similar to the 'Kitting' technique used in manual assembly, or whether the parts should be fed directly to the assembly station with the associated complexity, such as the example shown in Fig. 11,[5] and sorted there.

An example of centralising parts sorting, inspecting and loading into tubes is shown in Fig. 12. Here a Cybervision III System visually inspects keys for computer keyboards and loads them into storage tubes at the first station. At the second station the robot assembles them into keyboards. (See Reference 6 for more details). Both views have their advocates and the final answer is largely a function of the number of stations employed and the number of parts to be assembled.

Robotics at the present time is not standardised. This inhibits the growth of the field in a number of ways. One of these is that for lack of mechanical, electrical and signal standardisation no ready market exists for the large families of modular grippers which the industry needs and wants. As a result, grippers tend to be manufactured in low volume by and for each manufacturer. The costs are high and the art advances slowly.

Should this situation change, the day of quick change standardised grippers and fingers might arise with the resulting ease of specialising otherwise interchangeable assembly robotic stations. Also almost totally lacking are communication standards between robot controllers, and between controllers and other computerised equipment.

One hopeful advance is that the exchanging of geometric information from CAD/CAM systems to robotic systems is now becoming possible using the IGES II (Interactive Graphic Exchange Specifications) interface, now an American Standard supported by the National Bureau of Standards. The car door example previously shown in Figs. 9 and 10 also illustrates the experimental use of a CAD/CAM link between a Computervision Designer M and an Automatix Cybervision System. Here the fastening task is described and defined on the CAD/CAM system and is downloaded to the robotic assembly system for execution.

In conclusion, assembly tasks, ranging from miniaturised electronics and printed circuit boards or hybrid electronics to large assemblies such as automobile car doors or even riveting operations on airplane wings, loom around the corner. Economically attractive batch assembly is coming.

It is the author's opinion that within five years as the state-of-the-art moves assembly toward its full potential, robotic assembly will become the largest volume robot market.

REFERENCES

1. Westinghouse's 'Adaptable-Programmable Assembly System' (APAS) – in 'Assembly Robot Update', *Assembly Engineering*, pp. 42–45, April 1979.
2. M. Briot, Laboratoire d'Automatique (France). 'The Utilization of an "Artificial Skin" Sensor for the Identification of Solid Objects'. *Proceedings 9th International Symposium on Industrial Robots*. pp. 529–547, March 1979. Washington DC.
3. D. E. Whitney and J. L. Nevins, Charles Stark Draper Laboratory, Inc., Cambridge, MA. 'What is the Remote Centre Compliance (RCC) and What can It Do?' *Proceedings 9th International Symposium on Industrial Robots*, pp. 135–152, March 1979. Washington, DC.
4. Donald S. Seltzer, Charles Stark Draper Laboratory, Inc., Cambridge, MA. 'Tactile Sensory Feedback for Difficult Robot Tasks'. *Robots VI Conference Proceedings*, pp. 467–478. April 1982. Dearborn, Michigan.
5. J. C. Nevins and D. E. Whitney, 'Assembly Research', *Automatica*, Volume 16, pp. 595–615, February 1978. Photo taken from *Scientific American*, volume 238, no. 2, pp. 62–72. February 1978.
6. G. VanderBrug, D. Wilt and J. Davis, 'Robotic Assembly of Keycaps to Keyboard Arrays'. To be published in the *Robots 7 Conference Proceedings*, April 1983; Dearborn, Michigan.

The rise of the mini

Sarah Gardner of *The Industrial Robot* reports on the growth of small robots in education and for hobbies

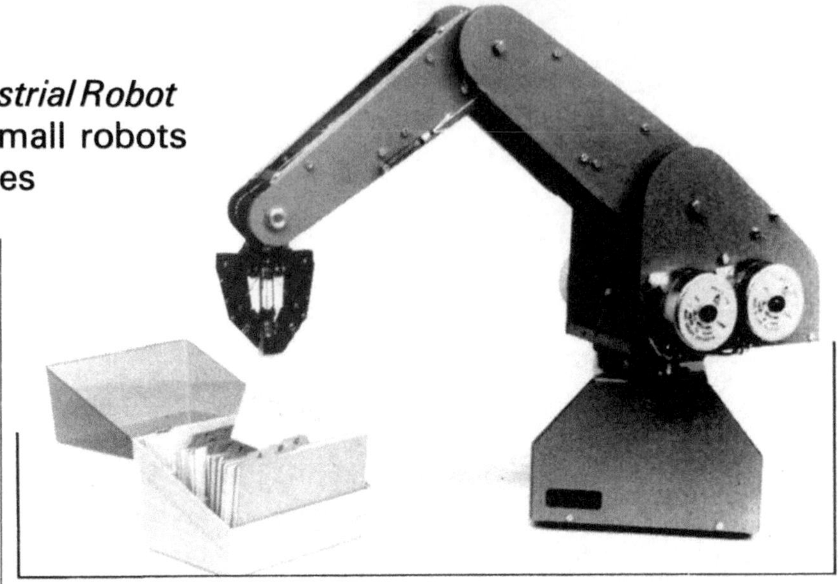

Fig. 1. The Armdroid from Colne Robotics.

SMALL ROBOTS for hobbies and educational purposes increasingly being termed micro robots – are a relatively new development compared to their larger industrial counterparts. They first began to appear on the market in 1980.

The main reason for their development was to enable industrialists and students to learn about robotics and control engineering without having to invest a lot of capital. Mini-robots are usually designed to imitate the much larger robots and therefore they are used to:

- ☐ Show how robots work.
- ☐ Show where robots can be used.
- ☐ Teach prospective users how robots can be programmed and subsequently utilised within industrial environments.
- ☐ Indicate their limitations.

A few companies have even shown an interest in using this type of robot for extremely light industrial applications where the expense of a larger robot, with a lifting capacity of a few kilograms, is prohibitive.

'Explosion in personal micro-robots'

The need to educate people in robotics is growing in both industry and education. Universities and colleges would rather let students develop their skills on a cheap robot with which they can safely experiment and learn the principles of robotics, before applying these skills in industry.

Colne Robotics produces an armdroid robotic arm – Armdroid 1 – which is available either in kit or assembled form. A set of instructions for both construction and operation is a part of the kit and contains full engineering drawings, electronic schematics and detailed software listing and description.

Severeal hundred Armdroids have already been delivered in 17 countries to universities, colleges, companies and individuals. It is used in a variety of educational, training and hobbyist roles.

The Armdroid robotic arm is a five-axis robot driven by stepper motors, it is a continuous path machine. The arm may be equipped with a three finger gripper (although modifications are available if required). All axes of motion can operate simultaneously providing point-to-point capability within the overall reach of 40cm. The technology is based upon both micro-electronics and electro-mechanical systems. Armdroid measures 6in x 9in x 12in.

It can be driven by most micro-computers and can be used as a handling device or alternatively as a computer peripheral. Advanced software has been written in machine code or in basic for several micro-computers to allow user-friendly programming. They are memory orientated and have a learning capability if required so that a robot will repeat a sequence which has been taught to it as many times as needed.

The product is priced between £300 and £500 depending upon the options taken. John Reekie, managing director of Colne Robotics, said: 'We believe that the explosion of interest in personal computers will be matched shortly by a similar growth in personal micro-robots, such as Armdroid. At Colne we are attempting to provide low-cost products which may be acquired for education, industrial feasibility studies, light production and so on'.

Colne is working on a larger 60cm arm (Armdroid II) which is capable of lifting 3kg and is expected to retail at about £1,000.

The teaching robot developed by Mitsubishi Electric is the Movemaster RM-101 micro-robot. It was developed in 1981 and first marketed in Japan in April, 1982. The main feature of the robot is the stepping motor control which can provide cheap and reliable movement with a minimum of mechanical complexity. The motors drive the joints of the robot through steel chains as opposed to strings as with some of the other robots. The grip has a pressure switch and the arm is built with steel box sections giving rigidity.

The robot can be programmed using its own set of single letter instruction. It can be interfaced to any personal computer with an 8 bit parallel output port. This is usually the printer output and instructions are given as print commands in BASIC. The single letter instructions are received by the Z80 microprocessor in the base of the micro-robot where they are recognised and decoded.

The RRM-101 has 6 axes of movement – five degrees of freedom. Three 'hands' are provided for widely different manipulative tasks. It can lift a weight of up to 500g and can remember 100 positions (this number can be increased to 600 if additional RAM is used).

Each of the stepping motors is positioned on the joint it controls, and this proximity enables repeatability of movements to be within 0.3mm. The speed of the micro-robot (7cm/s at tip of the hand) is suitable for trainees and is restricted so that mistakes can be checked in time to prevent any minor disasters.

The programming languages include BASIC, assembler, and MM-ROLY, a new robot-orientated language that further simplifies complex movements. Although the Movemaster RM-101 can be controlled by many of the widely

Fig. 2. The Mitsubishi Move Master.

available home computers, Mitsubishi Electric offers a system it has specially designed for the job, the MX6000 micro-computer.

The RM-101 costs in the region of £1,000. Developments are, however, underway to produce a cheaper and faster mini-robot.

Micrograsp, from Powertran Cybernetics, has an articulated arm jointed at shoulder, elbow and wrist positions. Control is by any computer with an expansion bus. However the ZX81 is suggested to be particularly suitable.

The entire arm rotates about the base and it has a motor driven gripper. Each of the arm movements is servo-controlled (there are position sensors feeding back information to the interface board where it is compared with the taught position and automatically takes corrective action). This servo action is independent of the computer, simplifying greatly the software to drive the robot and all programming is carried out with a small number of BASIC commands.

The interface board is memory mapped using only 64 bytes at any of 1024 switch selectable locations. The control signals required are the address bus, data bus, write and memory request or alternatively I/O device.

To enable a RAM pack as well as the interface board to be fitted to the ZX81 a three-way splitter board is available and the two edge connectors are then fitted to this.

To keep the price down and to increase its scope as an educational project the robot is supplied as a self-assembly kit in which all components are included, as follows:

☐ micrograsp robot kit with power supply £145.00.
☐ universal computer interface board kit £48.50.
☐ 23 + 23 way edge connector £2.50.
☐ ZX81 peripheral/RAM splitter board £3.00.

Sykes Instrumentation first started considering robots in 1979, when it required a small arm for a development that it was undertaking. At that time there were no small robots available in the U.K. The following year it contacted the American manufacturers of the Minimover 5, and subsequently became the UK and European distributors for this range of robots.

The Minimovers were an 'off-shoot' of the American space programme. The range uses what the Americans term 'tendon technology', i.e. they are electrically cable driven, using steel cable, with stepper motors controlling each axis of movement, and high tensile plastic gearing.

The MiniMover is a five-axis device with a parallel action gripper jaw rather than pinchers. The gripper is able to sense whether or not it is holding an object, and to judge the size of that object within $\pm 1/16$in.

The robot interfaces with Apple II, TRS-80 and Pet computers. It operates at low force levles so that it is forgiving of operator programming errors.

The TeachMover robotic arm incorporates all the MiniMover features but is self-contained in that it has an 'on-board' microprocessor. Microbot's teach control makes programming the TeachMover as simple as operating a hand-held calculator, but permits sophisticated routines. As many as 53 positions may be recorded in the unit's memory. It can be programmed to pause for up to 99 seconds or until it receives an external signal, and it can move at 16 different speeds.

Sykes has three main markets for the MiniMover range: education – universities and colleges; research and development in industry; and light assembly work.

As regards future developments it does not see any major changes in the two original robots in the range, but there may be a larger robot, capable of lifting upwards of 2kg, which could be used more on industrial assembly lines than in education.

Systems Control designed its first robot in January 1980. Each of the robots termed Smart-Arms, has six articulations. They are designed for continuous duty cycle and have been engineered specifically for simple maintenance.

'Robots with a variety of grippers'

Smart-Arms robots use dc servo motors for each articulation to give high torque/ inertia ratios, accurate positioning and the ability to measure drive torque for 'force feedback'. The servo command is via a pulse width modulation system to reduce noise pick-up in harsh electronic environments.

The robots may have one of a variety of grippers. For the research robots these are normally supplied with a standard electric gripper with the end effector in the wrist-hand-gripper configuration. While the industrial robots are normally supplied with an electric gripper configured as hand-wrist-gripper, this can be easily modified by the customer to be configured as hand-yaw-gripper.

The industrial robot can also be fitted with the pneumatic gripper in any configuration. For both the industrial and research robots it is possible to control from the interface and computer up to 8 servo controlled articulations. The maximum torque allowable for any end effector servo is 8kg/cm.

Smart-Arms I and R series robots are controlled by an Acorn computer. For the industrial version the computer is built into the electronics cabinet. In the research version the computer is separate for easy access for software development. The industrial robot has additional facilities including automatic start-up and battery backed RAM for sequence storage.

The interface is memory mapped so that it can be linked to most computers. The interface allows control of external processer as it has 8TTL compatible process inputs and 4 relay contact pairs for process outputs.

The software is written in high level BASIC allowing people with little experience in computer programming to operate the robot. The robot can perform a wide variety of tasks from 'pick-and-place' to 'search and learn'. The software can control associated equipment such as CNC machines and other robots.

The research robots are aimed at the same market as the educational robots but also at research and development departments of companies. The industrial robots are for use in light industry and for handling light dangerous material such as radio-active substances.

'All the features at a fraction of the price'

Systems Control is linking its industrial robots to CNC machines and CAD/CAM systems. It is currently selling Smart-Cells which are unit work cells, which can be linked together to form a complete flexible manufacturing system for small batch production.

The mechanical design of the robot has been aimed at maintaining all the desired mechanical features of the more expensive robots but at a fraction of the price. The design and development of mini-robots has not been restricted entirely to the UK, Japan or the USA.

A prototype model has recently been completed by an Israeli company. Sharnoa Robotics established in 1982, is a member of the large Israeli group Eisenberg. The GIL-1 model has a load capacity of 1kg and a maximum reach of

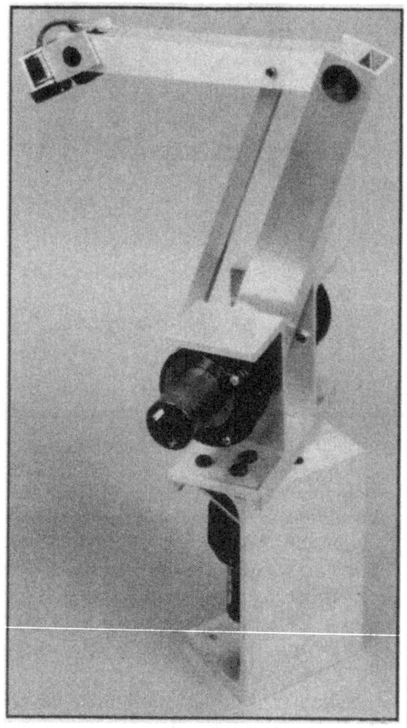

Fig. 3. Smart-Arms by Systems Control.

1m which makes it suitable for education, R & D and light industrial handling applications. This robot is the first of the new CIR 741 family of six-axis robots that Sharnoa intends to build this year.

Looking at things from the other side of the fence, one educational organisation is developing a robotics course is ORT (Organisation for Rehabilitation through Training). This body recognises the fact that within the current microprocessor technolgical revolution, robotics is one of the fastest growing fields. Consequently it is developing its robotics course with two main aims.

Firstly, the robot arm is seen as a means of visualising the capabilities of digital computers. This is most important in the area of computer literacy where, for example, young children can readily understand the fact that certain commands cause particular motions to occur. Thus in this context the robot is seen as a rather special form of computer peripheral device, which can translate computer events into 'real' events.

The second use is aimed at more senior students who are already studying engineering subjects, and robotics is viewed as a synthesis of many of these subjects such as microprocessors, control theory, feedback, drive mechanisms, programming and systems design.

ORT has purchased a Smart-Arm, and feels that it is an excellent educational tool, since it gives positive and demonstrable feedback to the program fed into its computer.

ORT is also looking at the Powertran range of hydraulic robots and a number of machines from the USA too. Here the object is to study as many different variations as possible, e.g. different drive mechanisms, control systems and methods of environmental feedback. It is felt that the Powertran robot has an advantage of being available in kit form and can, therefore, be used as a contruction project for schools.

CRAG (Cranfield's Robotics and Automation Group) with 35 researchers and 24 Master of Science (MSc) degree students is probably the largest group working on robots in Europe. It was the first intake of post-graduates to the MSc Robotics degree course who built CHARLIE (Cranfield's Handling Automation and Robot Learning Instruction Equipment). Keith Rathmill, Professor of Robotics and Automation at Cranfield and head of CRAG, stated that the development of CHARLIE was specifically an educational exercise for the robot application engineers being trained on the MSc programme.

'Inspiration from achievement rather than design'

Another important reason behind CHARLIE'S development was, given that industry works to deadlines and financial budgets, Rathmill felt it was most important that the students should demonstrate to themselves that they could build a robot inspite of the limitations in time (26 weeks) and money (£3,000).

Since its development several manufacturers have indicated an interest in developing CHARLIE commercially. However, although CRAG intend to make no secret of the design Rathmill would rather industry took inspiration from this achievement rather than the design.

Micro-robots on the market

	Armdroid	*Movemaster*	*Micrograsp*	*Minimover*	*SmartArm*	*CHARLIE*
Configuration	5 axes of revolution	6 axes	—	5 revolute axes and integral hand	6 axes	5 axes revolute design
Weight	8 lbs	10 Kg	—	—	—	—
Size	6in x 9in x 12in	—	—	—	—	—
Drive	6 stepper motors with open loop control	6 stepping motors	servo controlled	Electrical stepper motors – open loop control	—	DC servomotor
Controller	Any micro-computer with an 8 bit parallel port	MX-6000 Computer	—	Microcomputer provided by user	—	Has IEEE 488 interface therefore compatible with various microcomputers
Power requirements	15V 3 to 5 amps	115V ac 50/60 Hz 60W	—	12–14V, 4 amps DC	—	—
Gripper	3 fingers	3 different hands available	—	Parallel jaws	Various	—
Resolution	0.15in	—	—	0.25mm	—	—
Load capacity	10 oz	500g	—	—	Up to 200kg	1 Kg
Gripping force	5 lbs	—	—	3 lbs	—	—
Reach	17in	—	—	444mm	450mm 600mm 1000mm	1m working envelope
Programmable languages		BASIC, assembler M-ROLY	BASIC	ARMBASIC (specify interface)		CHAS
Cost	£300–£500	£1,000	see text	—	—	—

CHARLIE is a revolute type five-axis robot with dedicated microprocessor control and incorporates closed loop dc servo drives. It is capable of handling a 1kg load and operating with a 1m reach. The user communicates with the robot in a high level interactive language, CHAS. CHARLIE has an IEEE 4888 interface and is therefore compatible with various microcomputer systems.

Not only is CHARLIE now being used by students who take the MSc course as an important training aid, but developments and improvements will be made as future students gain experience with it. In fact, it is hoped that CHARLIE II will be ready by late 1983.

CRAG has not only developed a mini-robot but has also made use of the commercially available ones for training purposes although generally, Rathmill has been disappointed with the commercial micro-robots.

'Limitations greater than the true robot'

He sees their greatest fault as, not being like the larger industrial robots, and consequently they are misleading from the point of view of defining what a robot is. From demonstrations and experience of using these micro-robots he finds that many differ from their supposed larger counterparts in their programming and their limitations are far greater than the 'true' robot. So, far from teaching what a robot is, they could actually hinder and detract from any training programme that they are included in.

By no means are micro-robots expected to be as robust as their larger counterparts, but there should be strong similarities between the way they operate, are programmed and their limitations. One micro-robot that CRAG has made great use of, and been extremely satisfied with is the Smart-Arms. It is felt that the Smart Arms bears a close resemblance to an industrial robot, and that Systems Control has set out to make its products cost effective. Rathmill stated that 'At a cost of about £2,500 the Smart-Arm is

marvellous value for money'. CRAG have bought two Smart-Arms.

Technician colleges need real robots; few are in the happy position of being able to afford them. Hence there is a great need for micro-robots which accurately represent their industrial counterparts, and are more likely to fall within the financial range of these colleges. It is after all only with sufficiently trained technicians that industry can make full use of, and gain the real benefit of automation.

So far only a few technical colleges have bought mini-robots, so they are certainly behind industry in their use for training purposes. If financial resources permit the technical colleges could become a major market for mini-robots. But most of them in their present form will do more harm than good.

Apart from the types of training already mentioned that mini-robots can be used for, they can be used to demonstrate a group of robots built into a system as found in industrial situations. Companies have been using mini robots for some time in training their own staff to make them familiar with the new technology.

Schools and colleges are gradually becoming aware of robotics, but a certain amount of conservatism has to be overcome before a new subject is introduced to the curriculum. If the topic of robotics is analysed there is nothing particularly new about it. All its component topics, computing, control and mechanical engineering may already be included in the curriculum, and by teaching robotics it may even help students to grasp the intricacies of these subjects.

Microprocessors are in use in many schools already, so it seems that the use of micro-robots in schools would be a sensible development to see what computers can actually do. In Rathmill's opinion the leading schools will be using mini-robots within two years.

Just as microprocessors are being used in homes it seems possible that micro-robots will follow this trend and become another specialised hobby. Perhaps in the not-too-distant future such a hobby might gain a real boost if 'build your own Micro-robot' sets become available.

Rathmill feels that micro-robots as hobbies have a major role to play in accelerating the use of robots in British industry. Certainly the present trend indicates that as the big, commercial industrial robots prices, and numbers being sold, is decreasing, those for hobbies are increasing. It would therefore appear that the role of micro-robots is increasing in importance all the time.

So if present or future mini-robot manufacturers can move away from the present 'string-and-pulley-type' design, there is the possibility of a major market awaiting them.

Is robot technology safe?

Neil Percival of The Machine Tool Industry Research Association
considers the importance of and need for robot safety

WITH any new technology it is necessary to ask the question: Is the technology safe to use and to what lengths one should go to minimise the risks involved.

It is not justifiable to wait until fatalities occur and accident experience has been obtained and then to determine the level of safeguarding required. Yet it is easy to go to the other extreme with complex guarding and systems of work at unnecessarily high cost to both the manufacturer and user.

With robots the evidence so far is that they are dangerous. Although there have been very few fatalities (and none at all in the UK) many near misses and aberrant behaviour have been reported and the increasing use of robots suggests that practical guidance is required.

'Robots can present special dangers'

A Japanese survey[1] has shown that 36% of personnel working with robots have experienced an accident of some kind and that of these 8% have been injured. Further the survey indicated that the majority of accidents occur during programming, teaching and maintenance, and that there are relatively few accidents during normal operation. This pattern is also supported by Swedish[2] experience.

The difficulty with robots is that although they free persons from some of the more traditional hazards associated with machinery they can present special dangers. The major new hazard caused by the robot is its working envelope. Unpredictable action patterns, the robots ability to move in free space and the possibility of reconfiguration all distinguish it from other kinds of automated plant.

These movements, often at high speeds, can increase the risk of persons being struck by moving parts of the robot itself or by parts or tools carried or manipulated by the robot. Trapping points also can be caused by the movements of the robot particularly in relation to associated equipment such as work carriages or transfer mechanisms and trapping points can also be present on the robot itself.

From current experience it has been shown that there are six main sources of hazard:

- ☐ Control errors caused by faults within the control system, errors in software or electrical interference.
- ☐ Unauthorised access to robot enclosures.
- ☐ Human errors when working close to a robot, particularly during programming, teaching and maintenance.
- ☐ Electrical, hydraulic and pneumatic faults.
- ☐ Mechanical hazards from the parts or tools carried by the robot or by overloading, corrosion or fatigue.
- ☐ Environmental hazards such as dust, fume, radiation etc.

There are in fact no legal regulations in the UK specifically for industrial robots and therefore recommendations on safe operation and use have to rely on industry and government-formulated guidance.

The Health and Safety at Work Act 1974, is essentially an enabling act which places duties on manufacturers, suppliers and users to take appropriate safety measures but it does not tell them how these should be achieved. Where robots are installed in factories the Factories Act, 1961, is also applicable but this is also couched in general requirements such as, for example, that dangerous parts of machinery should be securely fenced. For particular robot applications then specific regulations such as the Highly Flammable Liquids Regulations and the Abrasive Wheel Regulations may also apply.

It was the lack of published recommendations on robot safety which led both the Health and Safety Executive (HSE) and the Machine Tool Industry Research Association (MTIRA) in the UK to consider the problems involved and, with the co-operation of robot manufacturers and users, to prepare guidance for the industry. These recommendations have now been published by the Machine Tool Trades Association as a booklet entitled 'Safeguarding Industrial Robots'[3] and have been well-received by robot users and manufacturers worldwide.

The difficulty in preparing guidance of this kind however is that practical experience of the day-to-day problems is limited and that, in general, solutions are based on applying well-proven safety methods and procedures to new technology. Whether there has been a

Fig. 1. Fences guard KUKA machines at Ford.

degree of 'over-kill' (no pun intended) to the safeguarding methods which are recommended remains to be seen and it could be that the guidance will need to be revised in the light of experience. Nevertheless the reaction so far to the recommendations has been excellent, particularly from the first-time user of robots.

'Safe guarding will usually be required'

The HSE recommends that in any new robot installation a risk assessment should be made. In 1981 the HSE produced a safety framework[4] and this in modified form has been used in the MTTA guidance and is currently forming the basis of tripartite talks with Germany and France on robot safety.

The framework proposed by the HSE suggests that each mode of operation of the robot should initially be considered, that is, normal operation, programming and maintenance. A hazard analysis is then carried out for each mode by considering first of all the behaviour of the robot as 'designed' and secondly by considering what is likely to happen if it behaves 'aberrantly'.

The latter is more difficult to determine as a hazard may only exist, for example, due to a control system malfunction. If the hazards which are identified are likely to lead to injury then this determines the standard of guarding required.

It could be that for very small robots or simple pick-and-place units guarding is not required. However for the majority of applications it is likely, as a result of the hazard analysis, that some form of guarding will be required with consideration being given to the need for access, the frequency with which access is required and the foreseeable risk and severity of injury.

Generally to allow safe access to the robot for teaching, fault diagnosis, repair and maintenance the guards will have to be carefully interlocked with the robot control system. In this way access is prevented when the automatic cycle is initiated.

The usual solution is to install a fence around the robot consisting of fixed and interlocked guards, although care has to be taken where access is required for teaching that the fencing does not cause additional trapping hazards. This is relatively straightforward where single robots are concerned. For example a typical application would be at Brook Motors of Huddersfield where a Unimate robot is used for feeding a diecasting machine and an extrusion machine and simple fixed and interlocked guards are used for guarding the accessible sections of the perimeter of the installation.

For lines of robots, however, a decision has to be made whether the whole line or individual stations are fenced off. On the British Leyland Metro body finish welding line the latter philosophy is adopted. This allows access to each robot and enables

Fig. 2. Fixed and interlocked guards are used to safeguard robot installations.

Fig. 3. Renault robots with fencing at Renault.

programming, maintenance and so on without necessarily restricting production.

The approach on the new Ford Sierra re-spot line, however, is different in that because of space limitations the whole of the line is guarded as one. On their other lines however where there are tool or jigging stages this over-all guarding concept is not practicable and individual guarding is necessary.

Alternatively or additionally, other forms of safeguarding can be used. These include physical stops within the robot to prevent the arm from moving outside the work area, trip devices such as photo-electric safety systems or pressure sensitive mats, or control system safeguards. The latter include interlocking built into the software of the robot control system itself.

Software interlocking, however, depends very much on the application and design and at the present stage of development is not considered to be sufficiently reliable to be used as a prime safeguard. For most applications, therefore, the HSE is recommending that, until more definitive data are available a separate hardware-based interlocking system should also be used.

Software interlocking is usually the prerogative of the robot manufacturer and built into the control system as opposed to most other safeguards which are generally installed by agreement between the manufacturer and customer and depend on the particular application.

The HSE recommends that consideration should be given by the robot manufacturer to safety at the design stage and where practicable robot mechanisms and controls should be of a fail safe design. Unnecessary trapping points should be eliminated and control systems should be ergonomically designed so that controls can be easy to identify and operate.

Some of the teach pendants on the first generation of robots were particularly confusing and it is encouraging to see that features such as automatic teach restrict speeds (not necessary for some applications), emergency stop, dead-man's control and hardware-based teach-restrict circuitry are now becoming more commonplace.

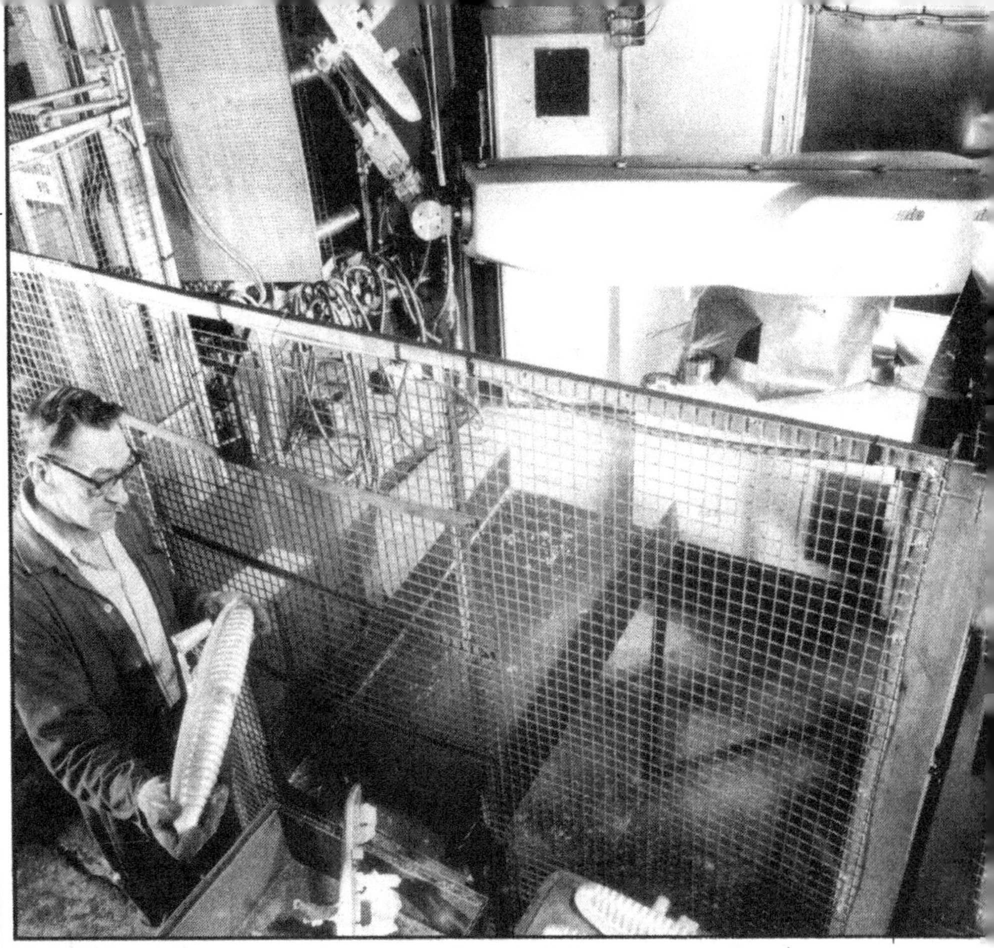

Fig. 4. A Unimate robot presents a casting to a photo-cell scanning device.

For the more complex robot applications where persons need to enter a robot enclosure for programming or maintenance it is also recommended that a safe system of work is used. Oral instructions, requests or promises are liable to be misheard, misinterpreted or forgotten and therefore correct documentation and access procedures are essential. The requirements for access have to be spelt out and recommended procedures for entry formulated to cover general access, fault diagnosis and maintenance, programming and emergency access.

In addition the most important safeguard is the training of all personnel involved in robot operation, programming and maintenance. They must be made aware of the operation of the control system and the consequences of malfunctions and the effects on the safety, both of themselves and others, of failure to follow the correct procedures.

Although the HSE and MTTA have published recommendations on robot safety there are as yet no official standards. Draft safety standards have been produced by Japan[1] and Germany[5] but most international work has been concerned with aspects of robots such as terminology, classification and performance.

Nevertheless the subject is not being neglected and committees are at work in the US, Japan and Germany. Work is also planned within ISO and possibly BSI and reference has already been made to the important tripartite talks of the UK, Germany and France and the work of the HSE Robot Safety Liaison Group.

As a result of the well-publicised fatality in Japan in 1981 a survey of accidents is being conducted in that country and a study group set up to produce a manual on the safe operation of robots.

All this activity is a measure of the concern being felt by robot manufacturers and users to ensure that robots are safe to use. Mention was made earlier to the few fatalities which have occurred but this should not give grounds for complacency as robots are being used in an increasing number of new complex applications.

The message is that safety must be given proper consideration at the early stages of design and installation. The UK should be congratulated on their initiative in producing guidance on good safeworking practice which hopefully will encourage the robot manufacturers and users to think 'safety' prior to installation and not as an afterthought.

References

1. Sugimoto N., Safety engineering on industrial robots and their draft standard safety requirements, Mechanical Division, Research Institute of Industrial Safety, Japan, 1980.
2. Carlsson J., Harms-Ringdahl L., et al., Industrial robots and accidents at work, TRITA-AOG-0004, Kungl. Tekniska Hogskolan (Royal Institute of Technology), Sweden, 1980.
3. Safeguarding industrial robots, Part 1: Basic principles, Machine Tool Trades Association, 1982.
4. Barrett R. J., Bell R., Hodson P.H., Planning for robot installation and maintenance: A safety framework, Proc. 4th British Robot Association Annual Conference, 1981.
5. Safety requirements for handling appliances and industrial robots, VDI 2853 (Draft), Verein Deutscher Ingenieure, West Germany, 1979.

Don't underestimate the need for training

Naqui Rizvi, marketing services manager, ASEA robotics division, outlines the training one manufacturer makes available

ADEQUATE user training is perhaps the single most important prerequisite for successful use of new technologies today. Within the user organisation training is necessary at several levels, from the decision makers and technologists to the shop floor operators and service personnel.

Training may be needed not only to provide a general understanding of the functions and pros and cons of the technology but to ensure the correct, safe and optimal operation of the equipment as intended by its designers.

When a new technology is being introduced it may happen that insufficient, improperly planned and carelessly run training could lead to operator alienation, malfunction complaints from the buyer and even personal accidents. All of these could in turn have serious repercussions on the supplier, as has been proved in practice.

'4,000 people have attended courses'

When dealing with advanced technologies like robotics it is not sufficient merely to provide the user with smaller or larger quantities of documentation for self-tuition so that he can gain confidence in the use and servicing of the equipment. Proper practical training is essential to provide a feel for the equipment and to transfer knowledge, and not least for overcoming the psychological barriers.

In robotics it has not only been a case of 'seeing is believing' but it is a matter of 'feeling' obtained through practical training involving the programming of the robot.

The importance of training in general and of practical training in particular has been accentuated even more in the case of robotics than of other comparable technologies for a number of reasons:

- The novelty of the technology and the enormous scope of its application
- Safety considerations – a robot is an 'open-ended' machine
- The presence of doubts and controversies resulting from the unfortunate comparisons with human beings.

Experience has shown that training programmes based on target group qualifications, practical application of the robot equipment, safety orientated instructions and pedagogically documented technical contents can give the buyer the information needed to justify his purchase, and instil self-confidence in shop-floor operators and service people.

During the decade 1973-83 ASEA will have trained more than 4,000 individuals worldwide who have attended the company's courses for operators, service men, production engineers and production managers.

User training was assigned high priority right from the beginning of the company's robotics operations in 1973.

The unit was set up as a separate function within the marketing department with responsibilities for scheduled and open course programmes, offered in the first place to buyers of robot systems for a separate fee.

The training operations were intended to break even and the policy of a separate course fee was applied to mark the importance and seriousness with which this essential customer service was regarded. At that time it constituted a clear deviation from the normal supplier practice of including training in the price of the equipment and, in certain cases, 'getting it over and done with at the eleventh hour'.

Increasing demand from users has led to investment by ASEA in a separate robot training centre at

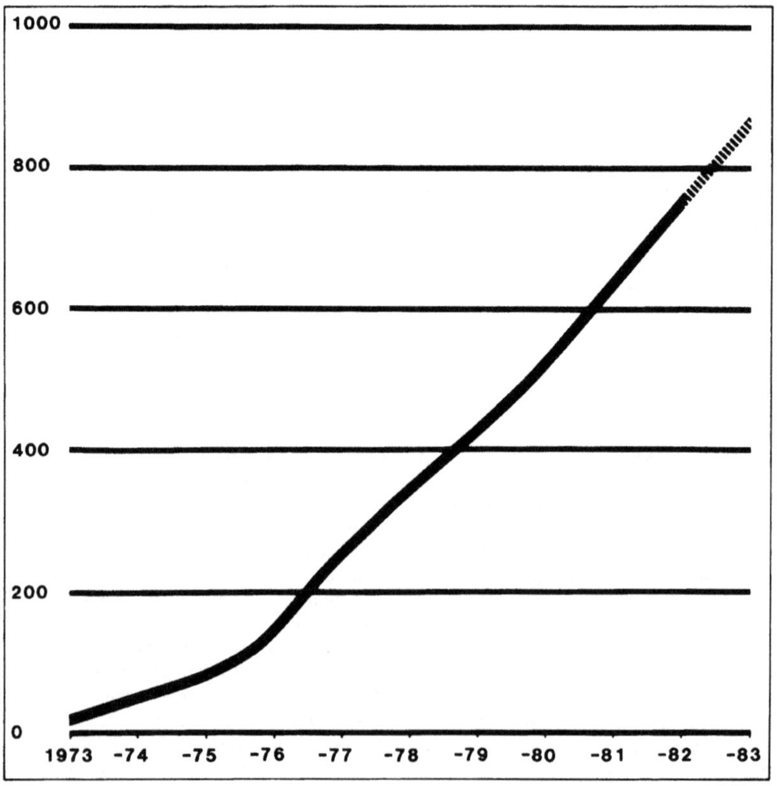

Fig. 1. A decade of training courses at ASEA with over 4,000 participants.

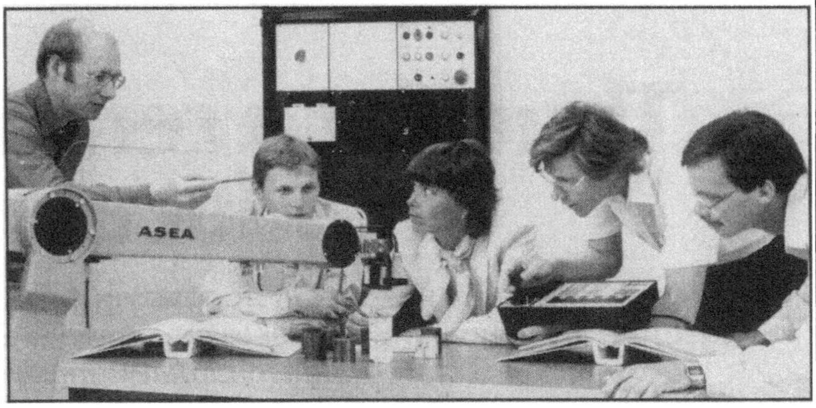

Fig. 2. User training is an integral part of the ASEA centre.

Västerås, Sweden. It was inaugurated on 1 April 1982 and complements the international training operations in robot centres in France, Germany, Italy, Japan, Spain, Sweden, the UK and the USA.

The training programme at the centre summarised here is typical of the courses run internationally and indeed of all good robot training practice.

The centre houses a workshop with 10 robot stations set up for different types of course. Each station is specially equipped with robot, worktable and auxiliary equipment such as a pen-holding gripper, index table and input/output signal simulation accessories.

Some stations have extra equipment to simulate real-life applications such as deburring, machine loading and advanced materials handling. Separate stations are equipped with fault-finding apparatus and aids, and mechanical parts are provided for dismantling and rebuilding robots for service engineer courses.

Three-day and five-day courses are available on most subjects, covering programming and operation, electronics servicing and mechanical servicing. There is also a five-day general course on project engineering and three- and five-day courses covering specific applications in project engineering.

The operation and service courses for the more advanced robots include 40-60% practical training. This limits the number that can be accepted to three to five trainees for each robot station, because each person must gain experience of programming.

On the theoretical side experience has shown that a total of 15-20 participants for a one-week course is about the ideal figure. This allows instructors to get to know every trainee within the short period in order to provide the right individual attention.

All the courses include complete documentation, varying from one to four well-filled files. Additional work has recently been put into tailoring the documentation to specific training needs with more pictorial descriptions, easier references and clearer texts. Special video films have also been produced to give basic functional orientation.

The general project engineering course is predominantly theoretical in the sense that the minimum of time is spent at the robot stations. However, the trainees have to work on a number of real-life application case studies in order to familiarise themselves with the problems involved.

Fig. 3. Courses for mechanics include drives and electronics.

These courses have a multi-level appeal to users. While managers obtain basic robotic knowledge to evaluate the technology, technologists are re-training themselves on the engineering aspects.

Project engineering courses devoted to a particular application have more practical content in terms of exercises at the application station. Problems solved are often actual cases selected from those submitted by the participants themselves.

Operator safety is an essential element of training. Even though accidents involving industrial robots are almost unheard of, the fact that the robot is an 'open' machine compared to, say, an NC lathe, makes robot operation a different matter. It is not difficult to envisage a situation in the future where a robot operator must be required to be a person certified through a course of proper training.

Some work is in progress on robotics standards and safety in a number of countries. If the key to start and operate a robot system is to be available only to *bona fide* persons, it makes sense that they must at least have participated in a well balanced operator training course.

Robotics is probably the most evolutionary of the modern production technologies. The last 10 years have set the scene for the evolution of different trades on the shop floor. A welder is becoming a robot programmer, a machine tool operator is learning to concentrate more on squeezing the maximum production from the machine-robot combination than tending the machine.

This change of trade requires training or, rather, re-training – training to utilise fully all the features of the technology at hand. It would be a pity not to, since the equipment can be considered relatively expensive.

One consequence of the predicted annual robot growth rate of 20-25% during the 1980s and the early 1990s is that both the decision-makers and the operation and maintenance personnel must continue to familiarise themselves with the pros and cons of robotics.

For the supplier the feedback of ideas from direct contact with users through training courses is a valuable base for improvements and developments. The importance of this information will certainly increase during the next decade as applications experience both widens and deepens.

ASEA is one company which is striving to maintain and improve the quality of training programmes on a worldwide basis. Apart from the training of teachers in, for example, speech and presentation techniques, further investments will be made in audio-visual aids, documentation and auxiliary equipment for practical exercises.

Time for a renaissance in education

Keith Rathmill, Professor of Robotics and Automation at Cranfield Institute of Technology discusses education and training of engineers in automated manufacturing systems.

WE HAVE NOW reached a milestone in the evolution and development of industry's needs for education and training in the field of automated manufacturing systems.

The emergence of the industrial robot as a now fully accepted part of industrial society is not only a herald of the new challenge of future manufacturing systems design, it is also a timely opportunity of great potential for re-establishing education and training on a more appropriate basis.

The robot combines a number of different disciplines and over the last decade or so it has been an increasing trend for educationalists to see the benefits of overcoming the political differences that exist between departments specialising in different disciplines. The emergence of the industrial robot not only provides a prime example of the need for inter-disciplinary training and education, its implications go much much further.

Certainly the robot in its stand-alone form clearly embodies a number of different disciplines. Prominent amongst those that spring to mind are mechanical engineering, electrical engineering, electronics, computer science, mathematics and not least production engineering.

'The multi-discipline nature of robotics'

Robots in the stand-alone case embody the requirements of mechanical design, drive systems technology, computer control, programming software and sensor technology. Given that industry is already seeking professional engineers to implement robot systems, clearly we are seeing a substantial reinforcement of the need for the multi-disciplinary education of robot specialists.

Traditional boundaries are being broken by the needs of both robot users and robot suppliers and this is one of the major problems retarding the supply of adequate numbers of appropriately trained technologists and technicians into these key industries. There is likely to be no one single best solution in the blend of training required for robotics specialists.

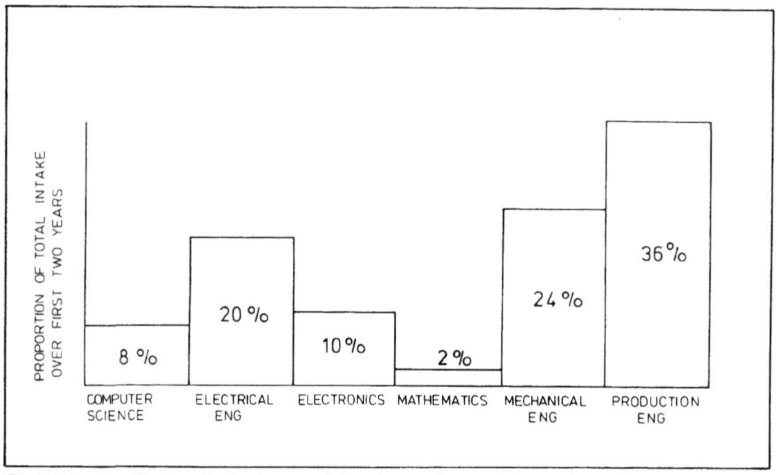

Fig. 1. The distribution of robotics specialists by previous training and experience.

If we look, for example, at the backgrounds of the students forming the first two years intake of the Industrial Robotics MSc programme at Cranfield (Fig. 1) we will see a typical distribution ranging over a number of disciplines and yet being characterised by a certain emphasis within the specific training programme. The multi-discipline nature of robotics as a subject area therefore raises a pressing need to escape from traditional boundaries both in terms of academic background of student intake and the traditional academic boundaries obstructing inter-disciplinary teaching and project development expertise.

One of the key mechanisms pressing for new programmes of training and education in the field of robotics is the relatively new problem of the rapid rate of change in the associated technology. There can be no doubt that the University system, for example, for all its many (and I believe there are many) strengths, is still largely organised around its traditional origins and Institutional structure possessing far too much inertia. This is not too surprising given the rate of technological change.

In the last three decades, it has been possible by and large for the established structures within the University system to cope with the relatively gradual changes in syllabus and training methods required of it. However, today the field of robotics is a fine example of our greater need to move ever more quickly in tailoring training requirements to a rapidly accelerating industrial technology. Quite simply our educational and training institutions must become far more responsive.

Educational inertia is not simply a matter of apathy, or indeed of outmoded practices such as tenure. There are major problems to be faced with the very need for the development of a National plan to develop centres of excellence where the increasingly high costs required to adequately support teaching and training in the required areas can be fostered.

With one or two exceptions training at the sharp end of an industrially-relevant technology has never been an easy job. However, there is no way in the context of the new technologies emerging, that course planning, lecture materials and laboratory programmes relevant today, could be expected to be adequate in say four or five years' time.

This fact places an increasing amount of responsibility upon training establishments to review thoroughly their operations on an annual basis. Indeed, the growing user-experience in robotics – for example in the assembly area – clearly will lead to major changes in the way that we teach this subject area within the next five years.

Robotic assembly, as yet in its infancy is clearly an area of major importance and one in which British industry

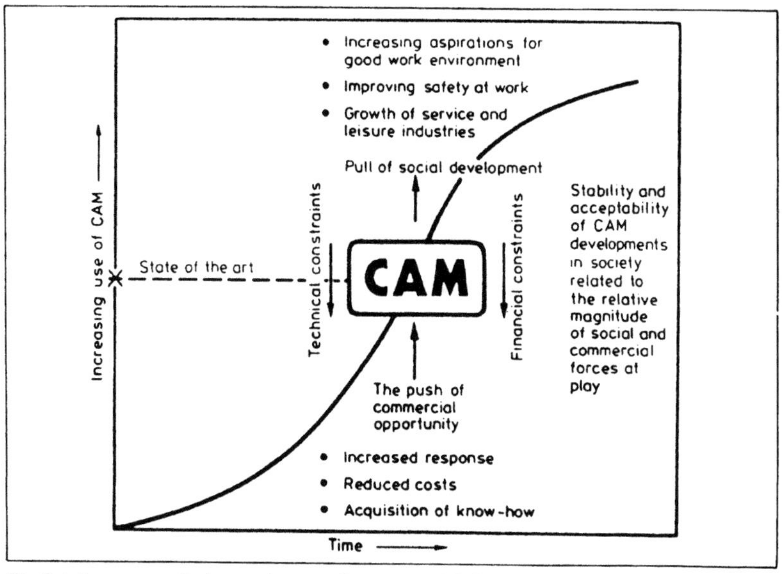

Fig. 2. Influences on the increasing use of computer aided manufacture.

desperately needs informed and expert personnel. One of the major roles of the specialist courses required will be to integrate a fast-moving industrial user experience base into the teaching and research programmes associated with the training of robot specialists.

Indeed, if we accept (Fig. 2) that due to the increasing effectiveness of world trading mechanisms there is a growing level of international competition pushing commercial interests, and that similarly the ever-increasing amount of research and development is removing the technological and cost constraints, the scene is being set in which an increasing rate of change will be limited only by man's capability to organise himself within industrial society to keep pace with that change. The technological change that we are facing within British industrial society is not the major challenge within itself, the major challenge is our ability to respond more rapidly than our competitors to this change. A key factor in the U.K.'s capability to keep pace with the rate of robot technology advancement is that of an adequate number of trained specialists.

For as long as I can remember, it has been a perennial area for discussion that more should be done to increase the status and appeal of the role of manufacturing technologists within society. However, quite apart from many historical and cultural problems within society working against this, there can be no doubt that for many years within industry there has been conflict between the shop floor technological requirements and the intellectual stimulus needs of the graduate.

'The key role of the manufacturing technologists'

No one with experience of industry would, I am sure, doubt the importance and the value of manufacturing industry to the Nation's economy and social well-being. However, many practising engineers in the last three decades have had to content themselves with a working environment and intellectual challenges somewhat less appealing than other graduates working in superficially more glamorous professions.

However, the picture that has greeted so many bright young engineers of a shop floor full of black art and mystery surrounding age-old workshop practices of which he knows nothing and feels totally alien, is rapidly fading before our eyes.

Replacing this culture is a rapidly diminishing labour force, the appearance of computer integrated manufacturing systems, as evidenced by shop floor data collection systems, the increasing practice of manual data input to CNC machine tools and shop and stock control systems with a distributed number of visual display units. Tool and cutter grinding are being replaced by disposable inserts, preset tooling and the major thrust new beginning to be felt in the onward sweep of CAD/CAM.

Clearly the use of industrial robots even as it is seen today in this new environment creates a picture of a completely different colour, and importantly offers all the solutions required to provide a genuine appeal for the key role of the manufacturing technologist in British industry today.

The challenges of the robot age engineer demand that he has expertise in a rapidly moving field of technology, that he is more than familiar with the practical realities associated with computer integrated manufacturing. Quite clearly we are seeking a new higher level of intellectual capability and expertise than has been required before.

However, I do believe the challenges we are faced with in the field of robotics are so clearly exciting and intellectually demanding that the subject will communicate its natural appeal and status to up and coming professionals and society at large.

The Industrial Robotics MSc Programme run by The Cranfield Robotics and Automation Group (CRAG) is now entering its third year of recruitment; was and still is the first and only degree programme of its kind specialising in Industrial Robotics. A cursory analysis of the students entering this programme confirms there is a much higher level of professional quality being attracted to this programme than to comparable conventional post graduate programmes. The quality factor is quite crucial for we can as an industrial nation only be as effective as the quality of our technologists and technicians permit. Importantly however it would appear from our experiences to date that there is a large body of extremely able young people within Britain that are seeking the opportunities which a specialist training programme in industrial robotics offers.

It is perhaps a personal hobby horse of mine, but I do believe the age of robotics comes at a time of increasing international industrial competition which requires new attitudes and perceptual capabilities of those involved.

Far behind us now are the days when the British Empire spanned the world, providing a captive market and the security for manufacturing industry that it clearly facilitated at the time. Tomorrow's – and indeed today's – manufacturing technologists need to be a lean, mean and committed set of people. It is important that a realistic awareness of the effects of the international competition threatening our industries is translated into a more aggressive than passive disposition.

The specialists we require need to be capable of thinking in total systems terms. We have left behind us the days of single machine replacement and the associated investment analysis being a central feature of manufacturing systems development. This is now being rapidly replaced by the design and financial/technical evaluation of dynamic batch producing inter-acting systems with varying degrees of computer integration.

The new breed of automated manufacturing systems developers are not seeking elemental or perfect solutions, rather they are looking for optimal or perhaps more accurately satisfying solutions, to complex multi-variable systems projects of increasingly high orders of capital cost.

'The need for strategic thinking'

Clearly, broad systems developments of major capital cost, affecting a wider scope of functions within the manufacturing organisation, require more than ever the ability for today's engineers to be capable of *strategic* thinking. An implicit challenge and requirement of robot automated manufacturing is the ability of manufacturing engineers to be capable of comprehending, analysing and communicating at the highest levels within company management on a strategic plain.

One could say that the attitudes, perception, orientation and general capability of engineers as briefly outlined above has been a longstanding requirement. To a certain extent this is true to the point that the lack of such capabilities have contributed to the longstanding problems of establishing a respectable number of manufacturing technologists in the key management posts directing British industry.

Notwithstanding this, the demands of robotics and its associated industrial automation are such that the success of U.K. Industry in the next decade will depend on the capability to think in positive, competitive and strategic terms. This capability needs training, and the robot specialists that we produce must be special in their greater understanding of this need.

Development of attitude and perspective requires a great attention to detail in the provision of the training environment and the type of instruction provided.

It is my belief that carrying out an exclusively abstract analysis of manufacturing systems smacks of a classical or traditional approach and one which does not succeed in engendering appropriate attitudes and motivations. A robotics specialist, therefore, requires a far more industrially oriented and realistically constrained project environment in which to develop his skills.

Without a doubt, industrial needs over the next decade do require bright, exceptionally capable, engineering specialists possessing initiative, drive and a strategic perception of their role. On a more immediate level, industrial needs at the present time are characterised more simply by an urgent demand for the volume supply of experienced robotics professionals. Generally, experience of industrial requirements in both robot user and supplier sides of industry, gained during the first three years of recruiting for CRAG's one year Industrial Robotics MSc programme is as follows.

Firstly, industry is aware that to move into the field either as a supplier or a user of industrial robotics, they need rather more than an enlightened or enthusiastic amateur. Put simply a bright enthusiastic engineer given the scope to run robotics projects within the company, it has been found, takes far too long to glean his information and experience.

There is therefore a tremendous de-

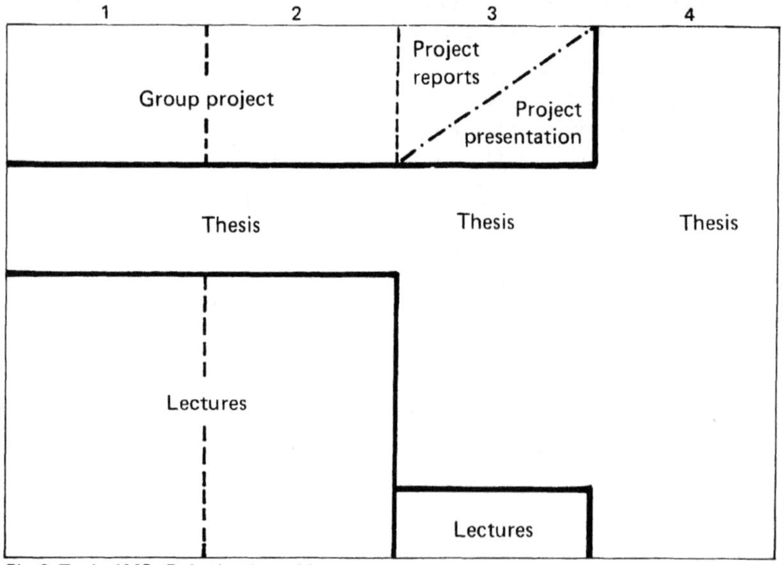

Fig. 3. Typical MSc Robotics timetable.

mand for experienced and fully trained engineers who can be fast off the mark in this field. This demand is reflected in labour turnover, a high profile of recruitment agency activity in the area and not least the excellent salaries and conditions being offered to attract and keep the few existing professionals who reasonably fit the bill.

Simply, companies recognise that they need to be fast off the mark in the field of robotics and that it is wise for them to pay for the expertise with which to do this.

CRAG's experience of the overwhelming demand for the robot specialists it has produced over the last two years indicates that the 25 or so robotics specialists emerging per annum from CRAG are entering a demand situation which could easily absorb at least 200 robotics specialists per annum and the demand is growing.

Given then the present state of a virtual vacuum of real robot expertise, a rapidly growing technology, and not least the fact that we are currently still in the grip of a very major recession, there would seem to be some serious grounds for predicting a serious supply/demand situation for robotics expertise here within the UK and internationally within the next five years.

Clearly, much of the good work going on throughout the UK in extending undergraduate courses to include options in industrial robotics of one kind or another are valauble and must be actively encouraged. However, a close study of the present industrial demand situation as briefly outlined above, clearly indicates that this alone will, by no means, satisfy industrial needs.

'Two year lead in training specialists'

At a professional technologist level therefore it seems highly appropriate to use the mechanism of an intensive one year specialist programme and given the intellectual calibre required of robotics specialists, it seems very appropriate that one should be using the well established academic identity of a Master of Science degree course for this purpose.

Some three years ago Cranfield, through its newly formed robotics and automation group CRAG, set out on an internationally leading endeavour to fulfil this role and is now recruiting its third successive intake of one year Industrial Robotics MSc specialists.

Although there has been a lot of international enquiry regarding this programme, demand from the UK has been such that it has not been found possible to satisfy requests from overseas students (including the United States) given the limited number of places available and of the numbers of highly qualified and able British applicants.

The CRAG team now believe that it, on behalf of the UK, has at least a two year international lead in the business of training robotics specialists. The CRAG programme is designed specifically to train robot application engineers and with vital support from the Science and Engineering Research Council (SERC), provides an essentially hands-on environment in which theory is complemented by a major commitment to practise during the year's training.

Fig. 3 depicts the breakdown of the students' commitments during what is an intensive year. The programme comprises three major components; firstly a lecture, laboratory, industrial speaker and industrial visit component. Secondly, a group project activity during which all students learn to organise themselves and inter-act in an intensive industrial systems development; and thirdly, the individual thesis project which usually is an industrial robot applications development of direct relevance to the student's sponsoring company.

The lecture programme which is very broad and offers many options contains training in numerical engineering, control, computing, microprocessor technology, interfacing, automation, robot technology, robot applications, computer aided design, flexible manufacturing systems, robot programming and robot sensory systems with particular emphasis on vision.

The students are encouraged to gain as much hands-on experience as possible of a wide variety of robots available within the laboratory environment. (Fig. 4).

Students are shown how to program the robots, are led by staff in considering the different capabilities of the various

Fig. 4. Hands-On Experience is a key requirement in training.

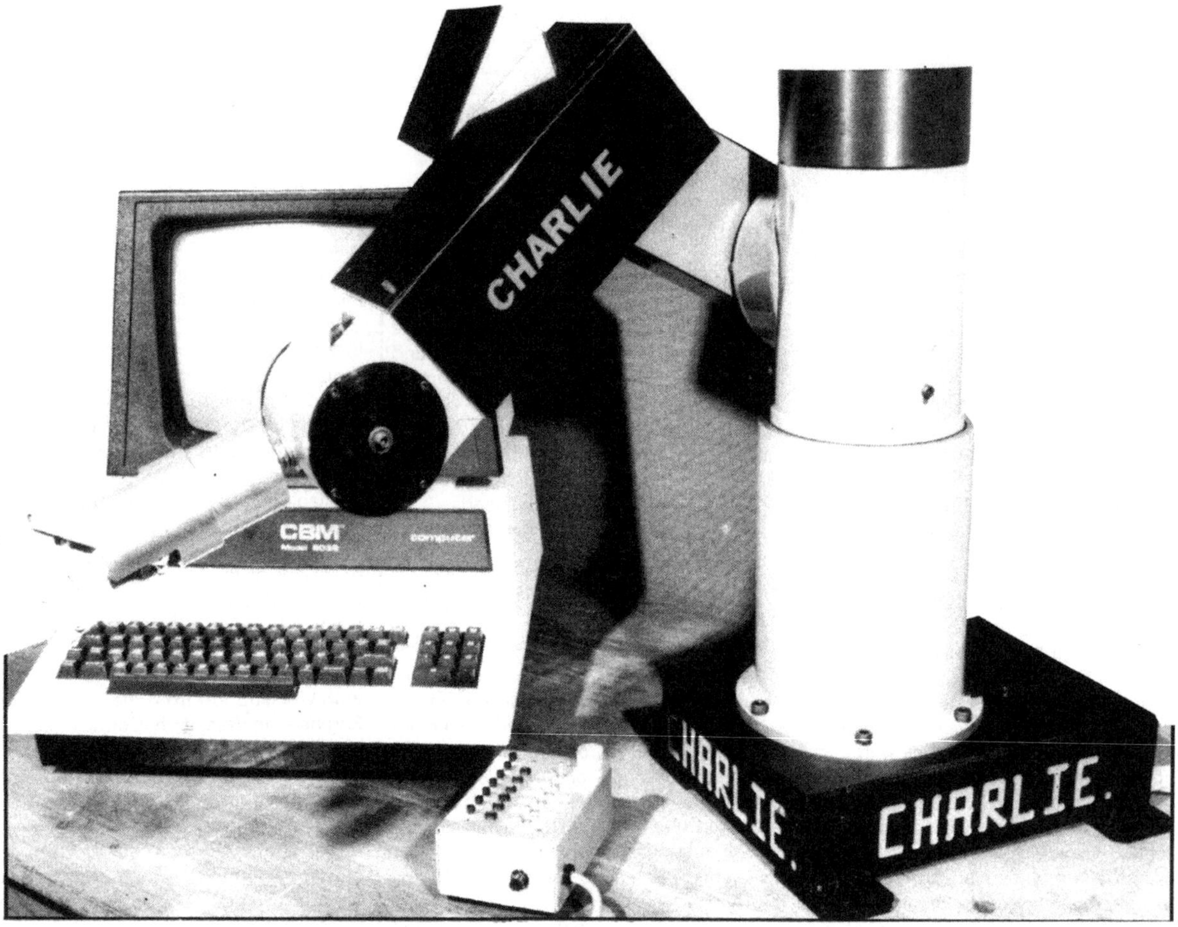

Fig. 5. Low cost robot project provides a good basis for group based training.

different forms of robot and the reasons for their varying performance. In addition, the students are encouraged to look into the structural design and software facilities associated with different types of robot and, after suitable instruction, to carry out simple tasks using robots of various kinds.

This hands-on experience not only begins to prepare the specialist for this industrial role but also complements and qualifies the theory he may receive in lectures and additionally prepares him for the intensive project work that he will undertake in his personal project and during his contribution to the group project later in the year.

The Group Project is designed not just to develop technical expertise within the students but, perhaps equally importantly, to develop their abilities in planning, co-ordinating and inter-acting in a dynamic fashion as a group on a major project.

With reference to Fig. 3, it can be seen that during the early part of their training, CRAG students are also subjected to three distinctly separate demands upon their time. This again is wholly intentional and is purposefully designed to train the specialists to think on their feet and be capable of planning and self discipline in their industrial environment. During the 26 weeks or so of part time work that went into the 1981/82 intake's Group Project, the group produced a protoype low cost 5 axis robot within a total budget of £3,000.

'Robotics specialist should not be under-valued'

The robot, named by the students as 'CHARLIE' (Fig. 5.) was designed, built and taken to a prototype development stage within the 26 weeks available, before being presented to 120 invited industrial guests. Whilst each year's Group Project is quite different, it was felt that whatever the merits of CHARLIE the students had benefited greatly from the experience of working together on such a demanding endeavour.

Subsequent to the completion of that Group Project, UK Prime Minister Margaret Thatcher made a generous gesture and invited a small team from CRAG to take CHARLIE to number 10 Downing Street where CHARLIE had the honour of shaking hands with The Prime Minister.

The individual thesis project carried out by the student can and should have a wide variety of forms, commonly however this is concerned with the development of flexible grippers for high variety parts handling applications, the application of robotics to a new production process such as metal stitching for example or the evaluation of a robot with vision sensing. (Fig. 6.)

Although CRAG's Industrial Robotics MSc Programme will continue to change and improve in an evolutionary manner for many years to come, there can be no doubt about its success and popularity with employers. Clearly there is a need for complementary courses to be mounted within the UK and it is encouraging to see the efforts of

Fig. 6. A student evaluates the Remek PAM-2 robot using a simple vision system.

Professor Paul Drazan at UWIST and Professor Tom Husband at Imperial College in this context. Whilst CRAG has developed its expertise in the industrially relevant training of robot applications engineers, it should not perhaps be forgotten that there are clearly important reasons for considering also the development of robot design engineers.

The robotics specialist at the technician level is clearly an emerging requirement and one which should not be under-valued. The interest of the Technician Education Council (TEC) in this field is to be valued and the efforts they are making in encouraging new courses within the Technical Colleges through their publication *The TEC News* and other channels is beginning to do good work in this direction.

Although the robotics revolution is as yet in its infancy, we are already beginning to experience the effects of a lack of robot service, robot test and maintenance technician skills in both robot supplier and user industries.

Those of us who believe in the future, I am sure, would agree that awareness and education must begin in the schools and in this context the BP Shell Build Yourself a Robot Competition can only be of immense value.

Similarly, The British Computer Society Voice Controlled Robot Competition supported by Honeywell and others having been designed specifically to cultivate and stimulate interest in Robotics within the technical colleges and polytechnics can only help encourage the work of those who are responsible for training our much neglected technician engineers.

Certainly the School of Engineering at The Hatfield Polytechnic and others who are now offering new robotics course options are leading in the necessary fast response to industry's pressing needs.

If we examine the present situation at a time of great unemployment and general recession, robotics specialists are already at a premium. Given the predictable rate of growth in this field it is evident that a restricted supply of high calibre professional and technician robot/engineering expertise is one of the greatest dangers threatening the growing effective use of robots in our industries.

Examining some of the deficiencies of existing training mechanisms and the new demands of flexible, reprogrammable automation generally, it is clear that in meeting the needs of industry we must sieze the present opportunity to create new specialist training courses. In so doing we can fulfill the potential of communicating a professional challenge of new scope and status to society at large.

It is clear to me that the UK has both the opportunity and the need to maintain an international lead in recognising and planning for the volume availability of high calibre robotics specialist skills. I believe that this opportunity is one of the most important affecting the future development of British manufacturing industry within the next decade.

Latter-day Luddites do not need sledge hammers

By Herman Rebhan, general secretary of the International Metalworkers Federation, Geneva

WORKERS do not consider robots as enemies. Why should they? In many cases robots take over work that is physically dangerous or boring. Struggling with the fumes and fire of a welding torch in a confined space is more punishment than pleasure and if a robot can take over such a task we should not complain.

But for trade unions – and with 14 million members in some 165 unions all over the world, my organisation, the International Metalworkers' Federation, has to relate to advanced Japanese, European and North American workplaces where robots are commonplace as well as to small engineering workshops in third world countries where robots only feature in science fiction movies – the problem of robots stretches well beyond the question of the taking over tedious work.

We have to ask ourselves several questions. Who benefits from the introduction of robots? Will displaced workers be found new jobs inside the company or will they be made redundant? Is there a regional or national retraining programme? Will overall employment diminish through natural wastage and non-replacement even if workers displaced by robots are kept on the payroll? Will the introduction of robots alter in a negative sense other patterns of work before or after that part of the production process where robots are introduced?

In short, will the arrival of robots benefit workers and the company they work for or are they being introduced simply to enhance profit at the expense of jobs and satisfaction in work?

'The source of profit is people'

I am not against profit, far from it, but the source of profit is people – ordinary people going out and purchasing cars, washing machines, video recorders, houses and holidays. As has been often remarked, the trouble with robots is that at the end of a shift they do not go out and buy anything. It is not just mass production that produced the remarkable post-1945 growth in the world economy but mass consumption. Robots, alas, consume nothing.

Both the IMF and its affiliate unions have been keeping a wary eye on robots for at least a decade. We see their arrival as part of the new industrial revolution, encompassing microchip technology and other aspects of new technology aimed at reducing the labour input in the manufacturing process.

We have noted alarmist reports but these need to be kept in perspective. For example, it is said that General Motors intends to have 14,000 robots in operation by 1990. These robots will replace workers who are now painting, welding, assembling, machine loading, palletising, inspecting, die casting and transferring parts.

It is said too that an average of five jobs are replaced by each robot, which means a reduction in employment of 60,000 in GM by 1990. This reads like a fearsome loss of employment. But right now there are 160,000 GM workers permanently laid off in North America because of the recession. Reagonomics, not robots, provides the more present and damaging threat to autoworkers jobs.

Last year we carried out a survey of IMF affiliates in the major manufacturing nations including the United States, Japan, Canada, Great Britain, West Germany, France, Denmark and Sweden. Contrary to some commentators who present unions as Luddites opposing the arrival of robots, our survey found that 'there are no IMF unions which resist totally the introduction of new technology as such, but most of them watch closely the development and introduction of new technologies.'

Instead, IMF unions are demanding joint consultation and, where it can be achieved, joint control over the introduction of robots. In 1979, the North American United Auto Workers signed an agreement with the Ford Motor Company regarding the impact of new technology on workers and the scope of the bargaining unit. The agreement provides for the establishment of the National Committee on Technological Progress between the two parties to deal with problems arising from the new technologies.

In Denmark, after several years of hard negotiation, a national agreement on technology was signed in the private sector between the Danish Employers Association and the national trade union centre, LO, in which the IMF affiliated Dansk Metal is very influential. The basis of the agreement is that 'there will be no new technology without consultation with workers.'

In Britain, the scientific and technical staffs union, ASTMS, has negotiated an agreement with Rolls-Royce setting up joint management-union machinery to ensure early discussion about the introduction of new technology.

The continuing introduction of robots will lead to four developments of importance to trade unions. Firstly, it is clear that in the long run the use of robots leads to reduced working time by increasing the productivity of labour. This may mean (and has meant) dismissals or non-replacement or, as we believe, it should lead to an increase in leisure time for employees, both workers and managers.

Secondly, we will see a transformation in the international division of labour. The advantage to manufacturing multi-nationals of newly industralising countries or developing countries has been the availability of wide-scale, cheap, and in terms of being non-unionised, docile labour. Once the costs of robots are amortised they are both cheap and docile so third world labour loses its advantages to the multi-nationals. It will make more sense to locate production closer to the best markets especially given current trends towards greater protectionism.

markets especially given current trends towards greater protectionism.

Thirdly, the introduction of robots changes business structures and jobs. The organisation of work, hierarchical structures, the division of labour and the quality of work, all face change. The nature of these changes is not determined by the arrival of robots as such, but by the relative strength and attitudes of unions and employers. However, robots will act as a catalyst for such changes.

'Pressing for reduced hours'

Fourthly, and this is tied more to the arrival of microelectronic forms of control in the manufacturing process, we are seeing structural changes in employment among sectors. Even within single companies we are witnessing shifts from the production sector towards the information and service sectors.

Arising from these changes, unions are advancing claims under three broad headings:

- Unions want to involve themselves in the political decision-making process connected with an economy's acceptance of robots and new technology. The medium and long-term implications of this new industrial revolution are such that they must be properly analysed at the national and international level. Social aspects must be given equal priority with the increase in productivity.
- Information, consultation and agreement jointly between workers and employers is a must. Make haste slowly should be our watchword. Massive investment in robotics in one year may prove disastrously wrong as new robots come onto the market which can do the task better and cheaper. Timely introductory information on new technology for all employees may prevent premature and possibly unjustified refusal of techniques. Vocational training and retraining is essential.
- The increase in productivity brought about by robots must be shared in terms of greater leisure time and maintenance of earnings. More than 40 years ago the United States moved onto a 40 hour week at a time when the rest of the advanced, industrialised world worked a 48 hour week or more.

American sceptics said that such a reduction of working time in relation to the rest of the world would ruin the US economy. The very opposite was the case. So unions today will continue to press for reduced working time as the corollary of the arrival of robots.

That, then, in brief outline is the trade union position faced with the arrival of robots in workplaces. But the severely aggravated and growing world recession has meant that trade unions are concentrating more on saving jobs, keeping people in work or encouraging political developments at governmental level which will turn round the world economy in the direction of growth. It was far easier to introduce automation in the expanding economy of the 1950s and 1960s when jobs were not so scarce as today.

To a certain extent, therefore, other issues, including the crucial one of new technology and the introduction of robots, do not command the same priority. But it would be a mistake to take the current defensiveness and low profile of trade unions as indicating a quiescent acceptance of whatever changes employers wish to introduce into workplaces.

It should be remembered that Luddism came about because of the introduction of machinery by *force majeure* without consultation with the workers involved. I do not see latter-day Luddites taking sledge-hammers to robots.

But the trade unions see the introduction of robots as part of the overall development of the world economy. The creation of wealth, which is what robots contribute to, must result in equitable distribution or the penalties will be high and perhaps a new Luddism will develop. To ensure that is not the case, consultation and prior agreement should be our watchwords in the workplace while at the national and international level, work for all, wealth for all, and growth for all should be our common aim.

The best-selling robot arc welding system in the world -the ESAB A30A

Today, there are more than 450 ESAB A30A robot arc welding stations in operation throughout the world. A considerable number, virtually unrivalled as far as we know.

The A30A success story depends to a great extent on the fact that it is so much more than just a robot – the A30A is **a complete arc welding station;** consisting of robot, control computer, welding equipment and workpiece positioner.

Thus, when you invest in an A30A station you obtain a well co-ordinated and highly efficient production unit, with first class components. This ensures profitable, top quality welding production.

If you would like some more information, we'd be only too happy to send you our new brochure "Teamwork – robot welding the ESAB way". Contact your ESAB representative or write to the address below.

See us on
Stand 1705
Automan '83 Exhibition 17th-20th May 1983
NEC Birmingham

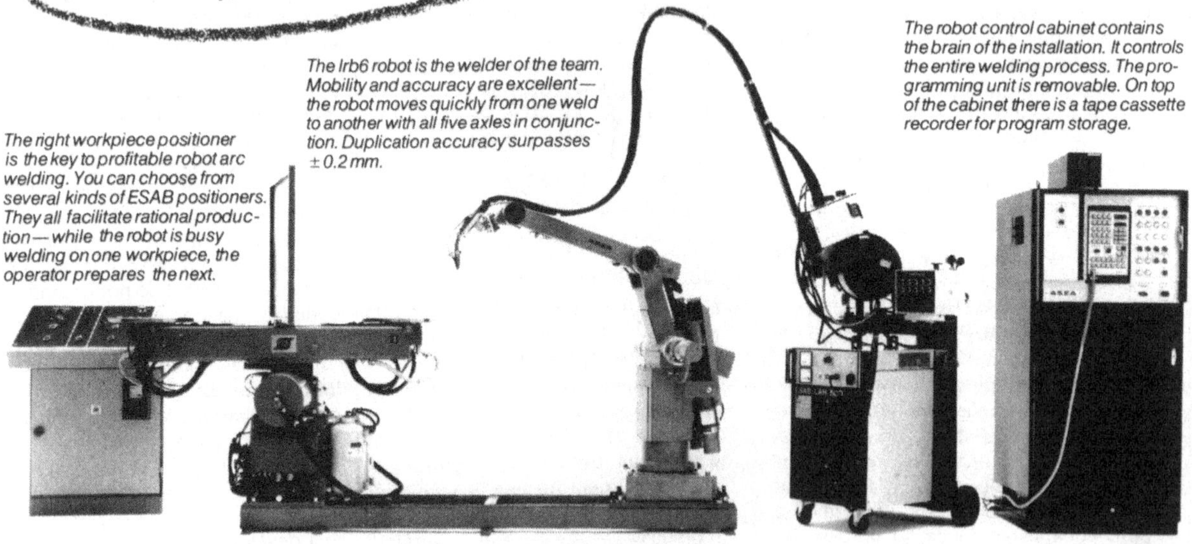

The robot is equipped with a very sophisticated power source and wire feed unit. There are four different versions; from Special Light Duty to Extra Heavy Duty.

The robot control cabinet contains the brain of the installation. It controls the entire welding process. The programming unit is removable. On top of the cabinet there is a tape cassette recorder for program storage.

The Irb6 robot is the welder of the team. Mobility and accuracy are excellent — the robot moves quickly from one weld to another with all five axles in conjunction. Duplication accuracy surpasses ± 0.2 mm.

The right workpiece positioner is the key to profitable robot arc welding. You can choose from several kinds of ESAB positioners. They all facilitate rational production — while the robot is busy welding on one workpiece, the operator prepares the next.

ESAB-LINCOLN Automation, Black Fan Road, Welwyn Garden City, Hertfordshire AL7 1QA Tel: 07073 24581 Telex: 268412 **ESAB**

IR4

Trade unions: Recognising the need for new technology

Anna Kochan of *The FMS Magazine* reviews union attitudes

THE clarity and detail with which trade unions have laid down their policies over the introduction of robots varies from country to country round the world. Sweden is one country which has an excellent reputation for good labour relations, and its Trade Union Confederation, or LO has clear views on the use of new technology.

Generally speaking, the LO recognises the need for investment in computers and electronics, of which the robot constitutes a small part, must increase if we are to offer security of jobs and welfare, writes Birgitta Frejhagen, the LO computerisation expert.

All robots existing in Sweden today carry out operations which were previously performed manually in bad working environments. The robots have not, however, been developed to solve the most striking problems in work environments because obviously the aim has been to achieve the highest possible profitability. But so far the need for profitable investment in robots has coincided with the need to get rid of certain hazardous jobs. That is why the results so far are good.

The 1981 LO congress adopted a programme of action, 'Trade Union Policy on Computerisation', which established the organisation's demands concerning the introduction of new technique at the workplace.

The programme stated, among other things, that computers and electronics shall be used to secure jobs and welfare, that the new techniques should be used to develop professional skills, and that computers should increase the possibilities of creating a democratic workplace organisation.

In order to achieve these goals, the LO is working on creating agreements, laws and instructions to enable its members to exert influence when new techniques are introduced at their workplaces.

In 1982, an agreement on co-determination was signed for the private labour market, which, among other things, stipulates that trade union organisations should have a right to participate during the whole developing process when a new technique is about to be introduced, that the employees at the workplace should get support through education, given in good time to enable them to carry out the jobs which are created when investment in a new technique is made, and that the trade union organisations have a right to engage their own consultants at the expense of the employer.

The LO also has a law on work environments which stipulates that a committee with representatives from both the employers and the employees, but containing a majority of employees, shall examine investments in robots with specific regard to the work environment aspects. State imperative mini-rules are also being elaborated for workplaces with robots.

In order to make use of the rights these laws and agreements provide, a broad consciousness and education among LO members is necessary. The LO has already started working on this. Up to last year, 30,000 LO members had participated in some form of trade union study circle or residential course on computerisation.

Trade Unions in other countries place security of employment in the same priority position as union involvement in the introduction of new technology. For example, the United Automobile Workers (UWA) in the US believes security is really the only basis on which automation can continue to be accepted in factories, according to Thomas Weekley, co-ordinator of the UAW skilled trades department.

The UAW skilled trades conference recently passed a resolution which is summarised as follows:

- Contracts must provide that workers be assured their income and employment security
- Advance discussion before the implementation of any type of automated equipment must take place
- The design used in new technology must be subject to bargaining in order to develop systems that help our members rather than degrade or place them in a secondary status
- Layoffs caused by the introduction of technological advance or change must be prohibited
- UAW craft workers will perform all the necessary skilled work
- There should be regular and periodic information meetings by us, as a Union, informing our members as to what is happening with new technology and keeping our local representatives updated on it
- Computers should not be allowed to be used for time study or disciplinary methods within the plant

In the future, according to Weekley, one area the UAW is going to have to address is lifetime employment guarantees. Companies point at the Japanese and their lower wages, but they omit to mention the fact that the Japanese are guaranteed a future with the companies they work for.

Britain's Trades Union Congress (TUC) is fully in favour of the introduction of new technology but is greatly concerned about the loss of jobs which can ensue if it is not managed properly.

'If the introduction of robots is mismanaged and takes place against a background of deflationary economic policies, and in the absence of positive manpower and industrial policies, the extra productivity could result in a permanent loss of a great many jobs. Such an approach is bound to produce an intense resistance to change on the part of the workers, points out Ken Graham, assistant general secretary of the TUC.

The necessary approach is set out in detail in the major TUC report on employment and technology which was adopted by its 1979 Congress and is more broadly developed in the recent Economic Reviews of 1982 and 1983.

The major themes of the TUC view are of the need for a reversal of current economic policies to one of expansion; the need for a viable British manufacturing capacity in the new technology areas; the need for plans to maintain, and, where possible, expand in the public and private sectors, employment opportunities; the development of opportunities for leisure time; and the need for a massive expansion of education/training.

In addition, the TUC rejects the far too widely held view that Britain should concentrate on exploiting the 'user versatility' of robots and their potential for increasing productivity, leaving the development of a manufacturing base to later.

The impact of new technology on working people

Dr Michael Cooley, Visiting Professor at the Technical Institute of Denmark looks at how robots may change people's lives

THOSE who initiate technological change at the conceptual level are invariably moved by the loftiest motives. That is a tradition of long standing.

Thus when he had completed the first mechanical calculating device the mathematician and philosopher Pascall said: 'I submit to the public a small machine of my own invention, by means of which you alone may, without any effort, perform all the operations of arithmetic and may be relieved of the work which has so often fatigued your spirit when you work with the counters or with the pen'.

At the level of implementation however the practice is frequently widely at variance with these laudible aspirations. Nonetheless the ethos persists!

'Improving the quality of life'

There is, therefore, a widespread belief that computerisation, automation and the use of robotic equipment will free human beings from routine, soul destroying, back breaking, hazardous tasks and leave them free to engage in more creative work. It is frequently added as a sort of occupational bonus that these developments will inevitably lead to shorter working weeks, longer holidae in more creative work. It is frequently added as a sort of occupational bonus that these developments will inevitably lead to shorter working weeks, longer holidays and more leisure time.

In a word that they will improve the quality of our lives in an all round way. Such an optimistic view is somewhat simplistic and does not accord with the experience of many trade unionists in recent years.

The traditional attitude of trade unions to new technology has been one of cautious optimism. That is to say, given certain safeguards, it could be accepted as essentially liberatory, human enhancing and progressive.[1, 2, 3]

During the 1970s there has however, been a discernible shift from this optimistic attitude. Initially, concerns related to straightforward questions of job and income security[4] and the provision of information on systems implications well in advance of implementation.[5] At that stage, employers seemed to be pursuing the first of the four industrial relations strategies defined by Ahlin and Svensson[6] namely, the Harmony Strategy.

This period was also characterised in some countries by support at the Corporation and State levels for joint humanisation of work programmes.[7] However, the economic recession and the growing problems of structural unemployment in the technologically advanced nations heralded a period, at first abrasive and gradually confrontational.

In Germany, disputes included issues of work cycle times[8] whilst in the US the Auto Workers' Union declared 'New Technology is a strikeable issue'. The anticipated, and in many cases promised, shorter working week, longer holidays and more leisure time – said to be one of the spin-offs of new technology, was not forthcoming although structural unemployment was growing. It is significant that the first strike in the German steel industry for 50 years centred on attempts to get a shorter working week.[10]

Although the metal-workers in Italy (FLM) have succeeded in some areas in negotiating a '150 hour sabbatical' as a form of creative job sharing,[11] the conflict on this issue is likely to be extended worlwide, as shown by the Australian unions.[12] Although it has long been recognised that it is difficult to specify precisely that component of unemployment which can be directly attributed to automation and computerisation,[13] it is nonetheless clear that technological change is resulting in significant labour displacements.

Nation states will increasingly have to recognise this reality and take appropriate steps. The new French government has promised reductions in the working week, and is also reported[14] to be considering combating structural unemployment by facilitating socially useful work as advocated by unions at Lucas Aerospace in the United Kingdom.[15]

These issues are still however merely at a discussional level and no infra-structure exists to cope with the growing problem of structural unemployment. Thus is it predicted that in the EEC countries by 1998 there will be some 20 million people out of work. Even in Japan 40% of the population are said 'To fear the effects of robotics and automation', and a recent fascinating report indicates that Japan may be heading for serious industrial problems.[16]

Robots are but one element in that totality which is referred to as 'the factory of the future'. Those who advocate widespread automation and increasing use of robotic devices do so on the grounds of efficiency. Yet in practice they succeed in squandering one of society's most precious assets which is the skill, the ingenuity, the creativity, the knowledge and the sheer enthusiasm of ordinary people.

The systems are invariably designed on the principles of so-called scientific management whose founder Taylor said on one occasion 'In my system the workman is told minutely what he is to do and how he is to do it and any improvement he makes upon the instructions given to him is disastrous to success'. This inevitably means that human intelligence and decision making is suppressed and marginalised by the systems.

The result is dramatic deskilling. With some of the numerically controlled systems in the US it is now suggested that the ideal workers are mentally retarded workers and a mental age of 12 is advocated.[17]

In this linear drive forward of technological change it is invariably forgotten that the most sophisticated systems have 'intelligence' levels which are trivial compared with the human mind. If one considers total information processing including pattern recognition the most advanced systems may have intelligent units which are to the order of 10^3. Yet a human being can bring to

Dr Michael Cooley.

bear something like 10^{14}.

Thus the widespread concern of trade unions about deskilling and, in many instances resistance to it, could force a creative rethinking of this rather blinkered technocratic development. In Britain, West Germany, Australia and the US trade unions are increasingly looking at the possibility of human centred systems. In these systems human skill and intelligence would be perceived to be an asset rather than a liability or as the systems designers would have it 'an uncertainty'.

'Subjective judgement of skilled workers'

Important research is now taking place into elements of flexible manufacturing systems which would respond to human intelligence and tacit knowledge rather than marginalise it. Professor Gossard's work at the MIT on analogical machine tools, in particular lathes, provides a possibility of re-examining this area anew.

More recently research with which the writer is associated at the University of Manchester Institute of Science and Technology with Professor Rosenbrock and his colleagues may provide a basis for systems in which the qualitative subjective judgement of skilled workers could be linked creatively with micro-processor based systems which would handle the quantitative elements. Thus a true human machine symbiosis may be achieved.

These machine tools would then comprise part of a flexible manufacturing cell served by robotic devices which would however by paced and controlled by the human being rather than the other way round. These possibilities are now the subject of intensive discussion in sections of the trade union movement. They could provide a framework in which unions would transcend that narrow economism which characterised them for too long and extend industrial relations such that the very design of the systems would itself become a subject of negotiation rather than merely the price at which deskilling and alienated equipment is introduced.

REFERENCES

1. G. Friedrichs, (Red) 'Qualität Des Lebens', Band 1-10 Europaische Verlagsanstalt, Frankfurt an Main, 1972.
2. Cf. (a) Statements in respect of Scandinavian and U.S. trade unions in Proc: 1st Conference on Industrial Robot Technology. Int. Fluidics Services Ltd., Bedford, U.K. 1973. (b) Various T.U.C. Reports 1965-1975.
3. S. D. Anderman, 'Trade Unions and Technological Change', Allen & Unwin Ltd., London, 1976. (Translation and edited version of L.O. research report, Stockholm 1966).
4. Anon 'Declaration for Full Employment and Income Security of Supervisory Staff', I.F.C.C.T.E., Geneva, 1974.
5. M. B. Sidell, 'Impact of Technological Change on Engineering, Drafting, Planning and Related Occupations'. Report at 41st Convention of I.F.P.T.E., Boston, 1974.
6. J. E. Ahlin and L. P. Svensson, 'New Technology in Mechanical Engineering: How can workers gain control?' Economic and Industrial Democracy. Sage, London & Berverly Hills, Vol. 1, 1980, pp. 487-521 (The Concept of the 4 strategies was first elaborated by Torsten Bjorkman).
7. G. Strauss-Fehlberg, 'Die Forderung nach Humanisierung der Arbeitswelt', Bund-Verlag, Koln, 1978.
8. R. Birkwald in 'Werktage Werden Besser', Europaische Verlaganstalt, Koln, 1978, p. 97. See also Stuttgarter Zeitung, 22.10.73.
9. H. Rhaiken, 'Numerical Control of Work: Workers and Automation in the Computer Age', *Radical America,* Vol. 13, No. 6, 1979.
10. S. Buschluter, 'German Steel Workers to Strike', *Guardian*, 27.11.78, p. 21.
11. S. Bodington, 'The 150 hour Agreement', *I.W.C. Bulletin,* Nottingham, UK, Feb. 1980.
12. (i) P. Nolan and C. Harvey, 'More Leisure Ours!' P.K.I.U. Publication, Sydney, 1980.
    (ii) Anon, '35 Hours: More Jobs – More Leisure', National Research Centre, A.M.W. & S.U., Sydney, 1980.
    (iii) Anon, 'Technology Means Fewer Jobs', C.B.O.A., Sydney, 1980.
13. S. Fabricant, 'Measurement of Technological Change', U.S. Department of Labor, Washington, 1965.
14. J. Palmer, 'French Boost for Alternative Work Pioneers', *The Guardian*, 24.7.81.
15. M. J. E. Cooley, 'Architect or Bee', Langley Technical Services, Slough, UK, 1980, pp. 63-74.
16. Galjaard, 'A Technology based Nation', Langley Technical Services Ltd, Slough, Berks, UK, 1982.
17. M. Cooley, op. cit.

The silent world of the undersea robot

Robert Wernli of the Naval Ocean Systems Center, San Diego, US, explores the use of robots undersea

IT is not often that one looks through the newspapers or magazines today without seeing an article on robots being used in the industrial environment. Whether considered a threat or blessing, these mechanical beings spark the interest of most people, and they are being readily accepted as co-workers and a complement to the work force. Most have also heard of space robots, be they satellites or planetary explorers, such as the Viking lander on Mars, which used its manipulator to perform work operations in an environment not yet accessible by man. Thus, land and space have long come under the robotic assault, and as one would logically expect, our last frontier, the ocean, has not been excluded from the revolution.

The ocean poses one of the most hostile environments that man can imagine; where extreme pressures, dynamic forces, corrosive attack, turbid water and other problems usually deal Mother Nature a winning hand. Therefore, it is no wonder that the ocean engineer would rather extend his presence into the ocean remotely, if possible, and remain topside in a warm, comfortable environment next to the coffee pot.

The undersea robot, generically called a Remotely Operated Vehicle (ROV), has actually been used by the United States Navy for nearly two decades for recovery of ordnance lost during testing on its coastal ranges. These vehicles, which usually have a power and control umbilical cable to the surface, use a claw or manipulator to routinely bring up hundreds of items.

Two non-routine recoveries performed by CURV III (Fig. 1) have been an atomic bomb off Palomares, Spain in 1965 and the manned submersible PICES III off Cork, Ireland in 1975 just prior to the exhaustion of the pilots air supply. One of the two men from the PICES quit piloting submersibles and opened his own ROV company, he had been quickly convinced of their invaluable aid in undersea operations.

The transfer of this technology moved rather slowly from the Navy into the ocean industrial community until the offshore oil boom in the mid-1970's. It became obvious that the smaller ROV's (Fig. 2) could provide valuable underwater observation without placing a costly team of divers in the water. They could also be used in situations too hazardous for divers, such as high sea states or near an undersea 'blow-out'.

'Initial resentment by divers'

Just as the workers who are replaced in industry by robots, there was an initial resentment by the divers to these mechanical intruders; however, they are becoming widely accepted by the community as they realise that they are a complement and not a total replacement. Diver assist vehicles are being developed that not only keep watch over the diver and carry his tools, but larger ones that the diver can ride on and manoeuvre during critical operations.

The technology is available to do much more with ROV's than industry is ready to accept. There is a lag time between development of systems and their accceptance by industry, as previously indicated by Navy vs. industrial application of ROV's. Although this period is decreasing, it will obviously remain.

For example, advanced work systems have been developed by the Navy, and now in industry, but they have not been widely accepted, at least when comparing them to the exponential rise in the number of observation type vehicles in use today. It is in the development of these more advanced systems that robotics technology, such as computer augmentation, will provide great strides in the advancement of the state-of-the-art.

The Navy's Work System Package (WSP) (Fig. 3) is an example of the type of work systems which will eventually be required to perform complex underwater tasks. These work systems would require, as did the WSP, multi-manipulator work suites which could exchange tools underwater and hold onto the object being worked on.

With this basic capability, the role of robotics or computer augmentation really begins to come into play. It allows the engineer to begin to optimise or produce a more efficient system.

Fig. 1. The US Navy's Cable-Controlled Underwater Recovery Vehicle (CURV III).

Fig. 2. SCORPI Inspection Vehicle.

Through the application of the computer, the following can be achieved:

- ☐ Preprogrammed tool exchange and manipulator movements
- ☐ Television systems programmed to follow the manipulator
- ☐ Simplified, more task oriented manipulator controllers
- ☐ Definition of planes or work surfaces
- ☐ Computer graphic overlays of the object being worked on
- ☐ Monitoring and limitation of manipulator and tool forces and manipulator movement.

'More advanced computer control needed'

These previous improvements will not only make a better system, but will relieve the operator of many of the fatiguing control tasks, allowing him to take a break while acting as only a 'supervisory operator'. The goal of supervisory control is to develop the proper integration of man, machine, and computer.

Eventually, the control console will incorporate voice interaction capabilities, allowing the operator and vehicle (or computer) to communicate using a sense previously underemphasised. The operator does not need all data at all times, and often does not need it presented visually. The computer then has the job to analyse the available data, and present the required data to the operator at the proper time in the most efficient manner. Thus, the computer carries much of the work load while the operator supervises the more critical aspects of the job.

The robotics applications just discussed address a 'real time' communications link. The goal for the future is to sever this link through the use of autonomous vehicles. These vehicles would require the use of more advanced computer control systems which woulsd direct it through its movements once it is launched. The computer can be programmed totally at the surface to fly the vehicle out, perform a mission, and return to the ship without further intervention from the operator.

An alternative is to use acoustic communication between the ship and the vehicle to call up the desired subroutine and redirect the path of the vehicle based upon the most recent events. Although this type of ROV requires its own on-board power supply, it would use less power and also eliminate the constraint of the umbilical cable, thus extending its capability. The cable, which provides the necessary link to allow the majority of today's ROV's to perform their job, is still the weak link in the system. They are being improved and a majority of cable related problems will be eliminated, however, the 'free-swimming' cableless ROV's of the future will be the next breakthrough.

Fig. 3. The Work Systems Package mounted on the Pontoon Implacement Vehicle.

The reality of these undersea robots will come about, not through the development of new advanced undersea component technology, but through the application of today's robotics technology which has been previously developed for industrial or space applications.

Adapting the position sensors, materials, computers, control systems, televisions, and other sensors to the harsh ocean environment is the problem. The advancement comes in the form of component adaptation and systems integration technology, which is no trivial task. Otherwise, we would probably have placed a man on the deep ocean floor prior to his walking on the moon.

The future should be kind to the ROV industry. The more they are accepted, the more industry will design equipment for them to maintain, and thus they will be advanced even more in their capabilities. This cycle should repeat, and similar to the application of robots to the assembly lines, they should become a common tool in the ocean. As the number of undersea robots increase, the problem of adapting components will hopefully decrease since there will be a market for support in this technology area at a component, rather than a system level. This trend will also increase the reliability of future systems along with reducing their cost.

Once we have provided the vehicle with a 'brain', it is a logical progression that we provide it with eyes. Following the successful application of vision systems to industrial robots, this technology will also be adapted to the ROV, providing those engineers working in the field of autonomous vehicles with their dream – a programmed vehicle that can go on a mission, find and recognise an object, perform an operation and return to report the results. It is not here yet, but it is certainly on the horizon. A horizon which will show the acceptance of ROV's as complementary tools by divers and users, similar to their industrial counterparts, and not as a clone of the human body or spirit.

Can the robot be the miner's friend?

By Professor Foster, head of the Mechanical Engineering Department, University of Aston in Birmingham, UK

BOTH hard rock mining and coaling are hazardous activities. Mechanisation in mines has proceeded steadily during the last 30 years, and more recently the National Coal Board has put considerable effort into the use of computers for data handling and overall mine management through the MINOS (Mine Operating System) project.

Nevertheless, men are still employed at or near to the coal face controlling the detailed operation of machinery. There is no reason why computers or microprocessors could not be used to raise the level of individual machine operation from mechanical power assistance to a degree of automatic control.

In a British coal mine the majority of the coal winning is from long wall faces at each end of which is an access tunnel or roadway, one containing the conveyor which carries the coal back to the mine shafts. The coal is cut by a shearer, with the roof supported by hydraulic pit-props.

Hydraulic pit-props can be made to automatically advance, and with the development of better sensors to detect the top and bottom of the seam, the operation of the shearer and pit-props could be almost fully automated. The roadways are advanced either by drilling and blasting or by continuous cutting of the rock by means of road heading machines.

As the coal face advances, surplus rock is packed behind the pit-props as they move forward and steel arches with shuttering are used to support the roadway. The machinery for advancing the roadways is still manually controlled and much of the work to line them is manual.

'Drilling jumbos are like large robots'

In some hard rock mining, for example, in the South African gold mines, the rock is very hard and the seams narrow, with the excavation of large amounts of rock being necessary to get at a relatively small amount of ore. Drilling and blasting are still carried out under manual control, but various types of continuous cutting are being introduced. In either case better control of machinery is a key factor for greater acceptance of the alternative techniques.

Drilling 'jumbos' which are used for drilling the holes prior to blasting look very like large robots and have several degrees of freedom of movement. In recent years drilling speeds have increased to the point where the time of operation of the boom itself is significant.

The boom, which is mounted on a movable chassis has to swing right, left and vertically up and down (that is, in spherical co-ordinates) in order to 'spot' the drill on the face. In addition, the carriage at the end of the boom has to have universal relative movement in order to align the drill hole. There may be more than one boom on a single machine, so that co-ordination is necessary.

The much heavier track-mounted road heading machine consists of a rotating cutter mounted on a single arm, moving again in spherical co-ordinates, but without the other movements of a drilling jumbo. The forces on each machine are high and hydraulic actuation is employed.

In order to carry out precise tasks with conventional manual valves in difficult co-ordinate systems, the operators of drilling jumbos and road heading machines and other mining machines have to acquire a high degree of skill. Several control systems are now being developed or are coming into service.

One drilling jumbo is now available from Norway which has been automated effectively for tunnelling, but has not yet been adapted to the electrical safety requirements of coalmines.

Fig. 1. Rotating head machine.

The Department of Mechanical Engineering at the University of Aston in Birmingham, under contract to the National Coal Board, has put microprocessor control on to the two main axes of a road heading machine, in the laboratory. This enables a tunnel profile to be put into memory via a changeable template and either the profile can be cleaned up or the whole sequence of cutting passes may be automatic.

Machines for coal mining are required to run on firesafe fluids, with intrinsically safe electric and electronic equipment, so that interfacing equipment used for normal robot applications is not usually acceptable. The machine at Aston employs a switching control strategy with poppet valves used for reliability, operating on dilute oil/water emulsion for safety.

The use of microprocessor control opens up more opportunities for machine development. With experience gained by both these lines of work, the completion of roadways could be made much more automatic than at present, for example continuous and automatic alignment of roadways or tunnels could be achieved, with health monitoring of the equipment becoming standard practice to give warning of impending failure.

'Labour is a high proportion of costs'

One application for conventional robots in the NCB which has been suggested is on the surface where filter presses are employed for dewatering the tailings, the final process in the washery plant. As many as 20 presses may be employed as a single washery, each having an assemblage of up to 150 individual plates. The tailings in suspension are fed into the gap between recessed plates, faced with a fine permeable cloth. After the application of a feed pressure for a period of time, the plates are separated and the resultant 'cake' of tailings falls on to the conveyor below; the plates are restored and the cycle starts again.

Labour is a high proportion of the running costs, and in spite of technical progress to improve the cycle, an average of about 20% of cake still fails to drop automatically and that which remains has to be removed manually. It is conceivable that robots could be used to replace some of the labour.

To conclude, there is a considerable opportunity in the mining industry for changing from mechanisation to automation and with the rapid change in the development of microprocessor technology, and the introduction of new interfacing devices, some of the earlier difficulties will be removed.

Thanks are due to Mr Tregelles, the Director of Mining Research and Development at the N.C.B., for permission to publish the article and support for the work at Aston, and particularly Mr J. Kibble for his enthusiastic guidance.

Fig. 2. Impression of coalface with hydraulic supports and tunnel excavated by roadheading machine.

ROBOTS IN SPACE

Ewald Heer of the California Institute of Technology, examines the use of robots in the hostile and remote atmosphere of space.

IN the production industry the primary motivations for introducing robots are productivity, quality and humanisation of the work place. In contrast, space and undersea operations (and other applications), require robots and remotely operated manipulators because of remoteness, danger, and inaccessibility to humans.

However, while applications for robots and manipulators are varied and manifold, many fundamental research problems are very similar, centering on vision sensors, and effector sensors, various artificial intelligence techniques, control system architectures, and peripheral interfaces.

To date, manipulator technology has been applied in the space arena in four instances. The first two were in the US Lunar Surveyor programme and in the USSR Lunar Exploration programme in the late 1960's. In the Surveyor programme a manipulator was used under remote control to dig a trench and perform other simple functions on the moon. In the USSR lunar sample return missions manipulators were used to acquire and deposit soil samples into the return capsule.

The third was in the US Mars Viking programme in 1976, when a computer controlled manipulator helped to perform a number of scientific experiments on the Martian surface. In 1982, the US Space Shuttle employed a Canadian-built 48ft long attached manipulator to unload payloads from its cargo bay and handle other objects in space (Fig. 1).

The actions of these manipulators are in the teleoperator mode or preprogrammed without sensory feedback. While the Surveyor manipulator is nonservo controlled, the latter two fall into the servo-controlled category. Developments are now under way to design future space manipulators with sensory feedback, that is as sensory robots freely manoeuverable in earth orbit (Fig. 2).

Planetary and Earth orbital spacecraft with functions of gathering scientific and utilitarian data and transmitting them to Earth are equipped with appropriate sensory devices, not only to acquire scientific information, but also to navigate, point accurately, and perform other programmed functions. These space robots also fall into the general class of sensory robots.

Future robots and manipulators in space may be put into four general application categories: (1) space exploration; (2) global services; (3) space industrialisation; and (4) space transportation.

Space exploration robots may be exploring space from Earth orbit like orbiting telescopes, or they may be planetary flyby and/or orbiting spacecraft like the Mariner and Pioneer families. They may be stationary landers with or without manipulators like the Surveyor and the Viking spacecraft, or they may be wheeled like the USSR Lunakhod and the proposed Mars rovers. Others may be penetrators, flyers, or balloons, and some may bring science samples back to Earth (Fig. 3).

All have in common that they can acquire scientific and engineering data using their sensors, process the data with their computers, plan and make decisions, and send some of the data back to Earth.

Some robots are, in addition, able to

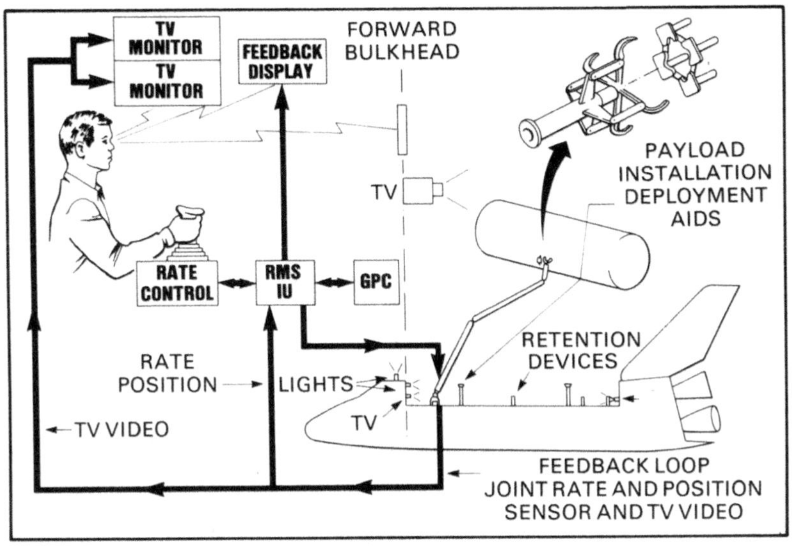

Fig. 1. How payloads can be unloaded from the shuttle bay.

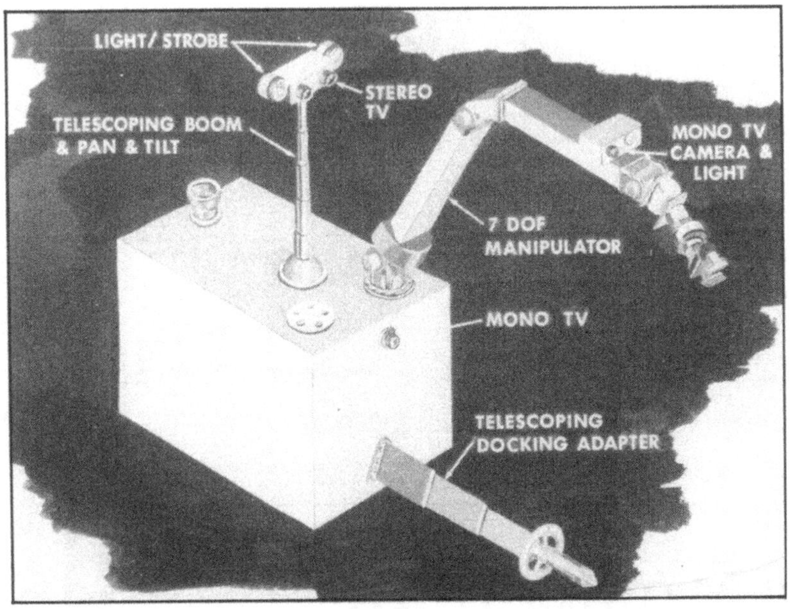

Fig. 2. Impression of a free flying remote manipulator unit.

propel themselves safely to different places and to use actuators, manipulators, and tools to acquire samples, prepare them, and perform in-situ experiments or bring them back to Earth.

It is clear that the spacecraft autonomous capabilities, including automated decision making and problem solving by the onboard computer system, will be major elements in the operation of these systems. The robot spacecraft must be able to implement sophisticated control strategies and manage its on board resources. It must also be able to assure self-maintenance of its functions and make the appropriate decisions toward achieving the assigned tasks.

Present and projected developments in machine intelligence suggest that in the future many of the presently ground-based data processing and information extraction functions can be performed autonomously on board the robot spacecraft. Only the useful information would be sent to the ground and distributed to the users. This would imply that the robot is able to make decisions on its own as to what data to retain and how to process them to provide the user with the desired information.

Space Industrialisation Systems. Such systems, including space utilisation and space transportation systems, require a broader spectrum of robotics and automation capabilities than those for space exploration and global services. The multitude of systems and widely varying activities envisioned in space until the end of this century will require the development of space robots and automation technologies on a broad scale. It is in this area that robot and manipulator technologies will have the greatest impact.

Large area systems, such as large space antennas, satellite power systems, and space stations, require large-scale and complex construction facilities in space. Relatively small systems, up to 100m in extent, may be deployable and can be transported into orbit with one Shuttle load, and handled with the Shuttle attached manipulator. For intermediate systems of several hundred metres extension, it becomes practical to shuttle the structural elements into space and assemble them on site, using, for example free-flying robots and teleoperated manipulators

Very large systems require heavy-lift launch vehicles, which will bring bulk material to a construction platform, where the structural components are manufactured using specialised automated machines (Fig. 4).

The structural elements will be handled by teleoperators or self-acting cranes and manipulators, which bring the components into place and join them. Free-flying robots will transport the structural entities between the Shuttle or the fabrication site and their final destination, and connect them. These operations require a sophisti-

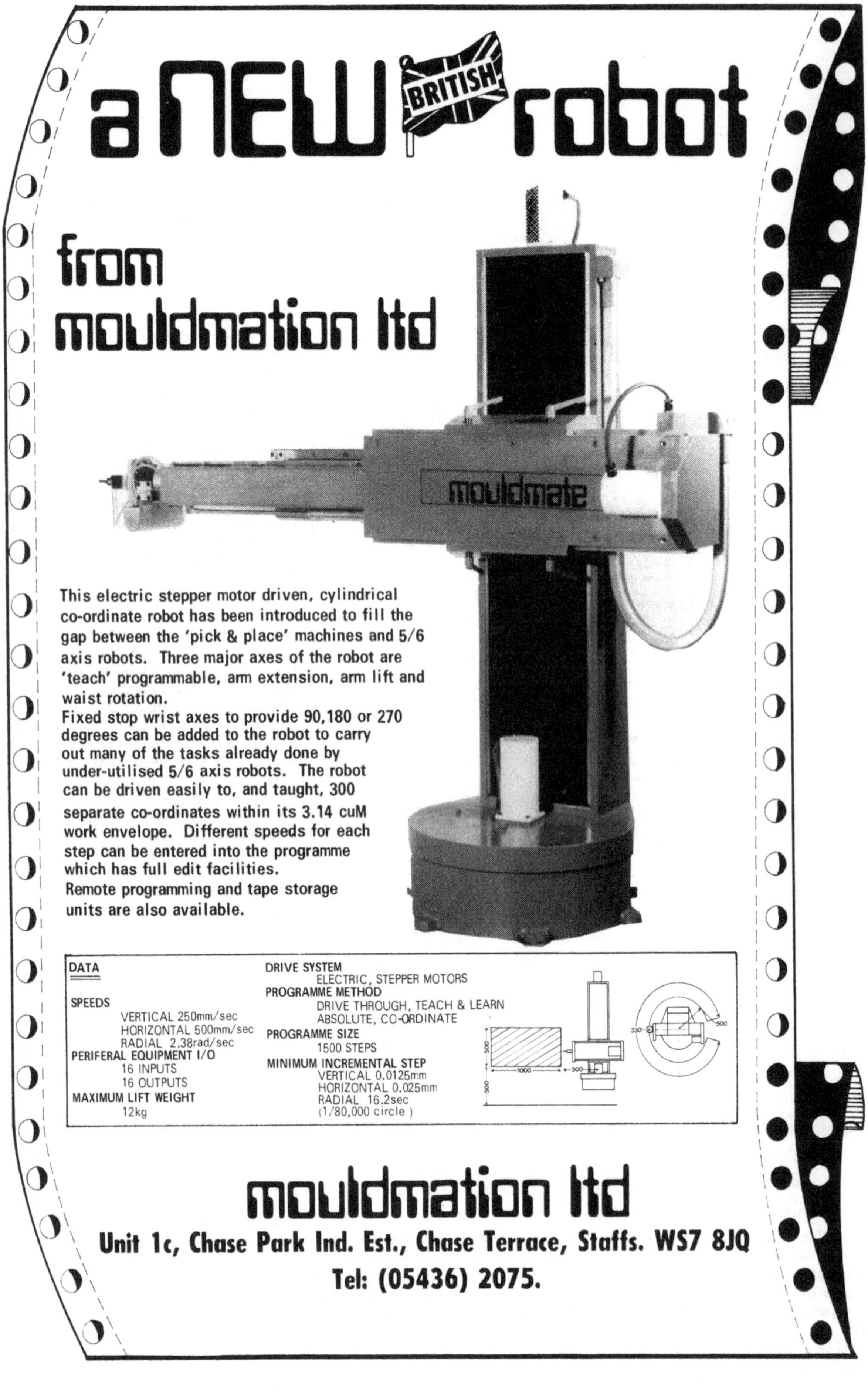
a NEW BRITISH robot
from
mouldmation ltd
mouldmate
This electric stepper motor driven, cylindrical co-ordinate robot has been introduced to fill the gap between the 'pick & place' machines and 5/6 axis robots. Three major axes of the robot are 'teach' programmable, arm extension, arm lift and waist rotation.
Fixed stop wrist axes to provide 90,180 or 270 degrees can be added to the robot to carry out many of the tasks already done by under-utilised 5/6 axis robots. The robot can be driven easily to, and taught, 300 separate co-ordinates within its 3.14 cuM work envelope. Different speeds for each step can be entered into the programme which has full edit facilities.
Remote programming and tape storage units are also available.
DATA
SPEEDS
VERTICAL 250mm/sec
HORIZONTAL 500mm/sec
RADIAL 2.38rad/sec
PERIFERAL EQUIPMENT I/O
16 INPUTS
16 OUTPUTS
MAXIMUM LIFT WEIGHT
12kg
DRIVE SYSTEM
ELECTRIC, STEPPER MOTORS
PROGRAMME METHOD
DRIVE THROUGH, TEACH & LEARN
ABSOLUTE, CO-ORDINATE
PROGRAMME SIZE
1500 STEPS
MINIMUM INCREMENTAL STEP
VERTICAL 0.0125mm
HORIZONTAL 0.025mm
RADIAL 16.2sec
(1/80,000 circle)
500
1000
500
500
330°
500
mouldmation ltd
Unit 1c, Chase Park Ind. Est., Chase Terrace, Staffs. WS7 8JQ
Tel: (05436) 2075.

cated general-purpose handling capability.

In addition to transporting structural elements, the robot must have manipulators to handle them, and work with them and on them. Large structural subsystems must be moved from place to place and attached to each other. This usually requires a rendezvous, stationkeeping, and docking operations at several points simultaneously, and with high precision. These robot systems could be controlled remotely like teleoperator devices, or they could be under supervisory control with intermittent human operator involvement.

Maintenance and Rescue. After a system has been constructed, its subsequent operation will require service functions that should be performed by free-flying robots or by robots attached to the structure. The functions that such a robot should be able to perform include calibration, checkout, data retrieval, re-supply, maintenance, repair, replacement of parts, cargo and crew transfer, and recovery of spacecraft.

During and after construction, there should be a robot on stand-by for rescue operations. An astronaut drifting into space could be brought back by a free-flying robot. Such devices could also be on stand-by alert on the ground.

The delivery systems for these rescue robots need not be man-operated. They can deliver expendable life support systems or encapsulate the astronaut in a life-support environment for return to a shuttle, space station, or Earth. They could also perform first-aid functions.

Materials Processing. Utilisation of space includes processing of materials in space. Space materials processing requires additional sophisticated technology, including remotely operated automated machinery and robotic devices. Basic types of processes envisioned include solidification of melts without convection or sedimentation, processing of molten samples without containers, diffusion in liquids and vapors, and electrophoretic separation of biological substances.

It is expected that techniques will also be developed for the acquisition and conversion of raw space materials to forms useful for space system components and for other space and terrestrial applications. A first step in this process calls for a lunar base.

Fig. 3. A Mars surface scientific sample return facility.

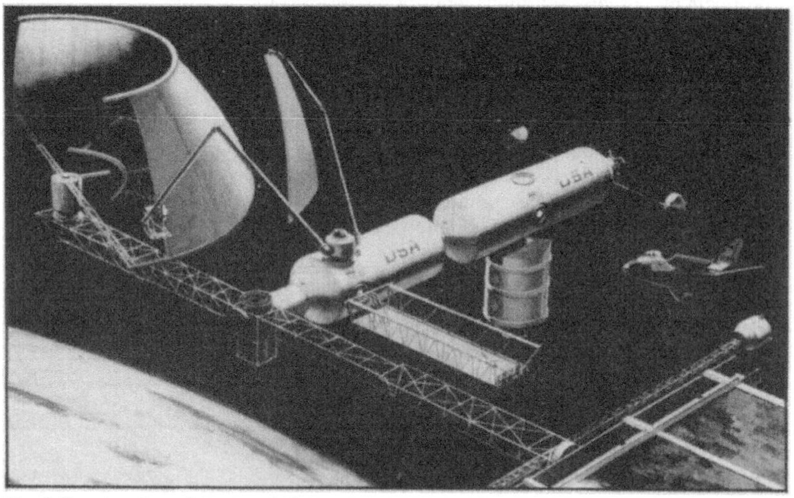

Fig. 4. Construction of a space station.

Fig. 5. Automated material processors are serviced by robot vehicles.

After a lunar surface survey with robot (rover) vehicles, an automated precursor processor system could be placed on the Moon. This system would collect solar energy and use it in experimental, automated physical/chemical processes for extracting volatiles, oxygen, metals, and glass from lunar soil delivered by automated rovers (Fig. 5). The products would be stored, slowly building up stockpiles in preparation for a lunar base. The lunar base would be constructed using automated equipment and robots similarly as in Earth orbit.

After construction, general-purpose robot devices would be necessary for maintenance and repair operations. In addition, the lunar base would use all types of industrial automation (qualified for operation in space) that are generally used on Earth for similar tasks.

Japan's robot industry: Where does the future lie?

Kanji Yonemoto, executive director of the Japan Industrial Robot Association, looks at future prospects

THE PRACTICAL use of industrial robots in Japan over the past 10 years or so has produced a number of economic and social advantages. They include an improvement in productivity, greater humanisation of working life, prevention of occupational and industrial accidents, improvement of product quality, and early return on capital investments.

These economic and social effects stemmed from the fact that the industrial robots are flexible and versatile. This has led to the automation of multi-product small-batch production, and changed the production system from a man-machine system to a man-robot-machine system. This change means that people have been released from dangerous, unfavourable work.

Efforts have recently been made to popularise the use of industrial robots in Japan through policies implemented in 1980:

- □ Establishment of a leasing system and of leasing companies designed to make industrial robots more attractive to small and medium sized companies.
- □ Special finance from the Small Business Finance Corporation and the People's Finance Corporation to small and medium companies for the introduction of industrial robots designed to ensure workers' safety.
- □ Establishment of a special depreciation system for high-performance computer-controlled industrial robots.
- □ Application of loans and leasing programmes to industrial robots by local governments to help small firms to modernise their equipment.

The use of industrial robots in Japan advanced throughout the 1970's, with a total of 57,000 machines in use at the end of 1979. However, it is predicted that substantial popularization of industrial robots in Japan will occur after the 1980's (see Figs. 1 and 2). It is predicted that demand for industrial robots in the manufacturing industries in 1985 and 1991 will be approximately Yen 290 billion and Yen 520 billion, respectively.

Furthermore, it is expected there will be substantial future demand for industrial robots in non-manufacturing such as nuclear energy, ocean development, and public works and construction, as well as in service trades such as medical services and transportation.

Table 1. Classification and Definitions of Industrial Robots
(Japanese Industrial Standards JIS B 0134–1979)

Classification by input information and teaching method	
Manual manipulator	A manipulator that is directly operated by a man.
Sequence robot	A manipulator, the working step of which operates sequentially in compliance with preset procedures, conditions and positions.
Fixed sequence robot	A sequence robot as defined above, for which the preset information cannot be easily changed.
Variable sequence robot	A sequence robot as defined above, for which the preset information can be easily changed.
Playback robot	A manipulator that can repeat any operation after being instructed by a man.
Numerically controlled robot	A manipulator that can execute the commanded operation in compliance with the numerically loaded working information on e.g. position, sequence and conditions.
Intelligent robot	A robot that can determine its own actions through its sensing and recognitive abilities.
Classification by motion	
Cylindrical coordinate robot	A manipulator which moves primarily in the cylindrical coordinate system.
Polar coordinate robot	A manipulator which moves primarily in the polar coordinate system.
Cartesian coordinate robot	A manipulator which moves primarily in the cartesian coordinate system.
Articulated robot	A manipulator which consists primarily of an articulated arm, that is, revolute joints.
Note: Manipulator:	A device for handling objects as desired. Without touching with the hands and it has more than two of the motional capabilities such as revolution, out-in, up-down, right left travelling, swinging or bending, so that it can spatially transport an object by holding, adhering to, and so on.
Robot:	A robot is defined as a mechanical system which has flexible motion functions analogous to the motion functions of living organisms or combines such motion functions with intelligent functions, and which acts in response to the human will. In this context, intelligent functions mean the ability to perform at least one of the following: judgement, recognition, adaptation or learning.

Industrial robots at present are those devices which provide versatile and flexible moving functions similar to those of human limbs and enjoy intellectual functions through their sensing and recognition capabilities. The Terminology Standardisation Committee (TSC) of the Japan Industrial Robot Association (JIRA) worked out the classification and definitions of industrial robots in 1977 as shown in Table 1, and this was officially incorporated into the Japanese Industrial Standards in February 1979.

Along with the anticipated technological progress, however, there may come into being those robots which can perform such motional functions as for animals like snakes and crabs. Hence the classification and definitions of industrial robots will need to be modified as technology continues to progress.

There have been a number of factors affecting the diffusion of industrial robots throughout the country.

□ First, the development and practical application of industrial robots were stimulated by a serious labour shortage in late 1960's, a decade which witnessed the nation's highest economic growth. GNP showed an average 12% growth annually during this period, whilst the shortage of labour, particularly skilled labour, became more acute and widespread, reaching as many as 1,800,00 in 1965.

□ Second, the first oil crisis in October 1973 forced Japan to change direction to low economic growth. As a result, the strain on the labour market eased to a remarkable degree, but the country's highest oil and other natural resources bills pushed up commodity prices drastically and hence labour costs. To fight against such inflationary effects required a dramatic improvement in productivity, encouraging private

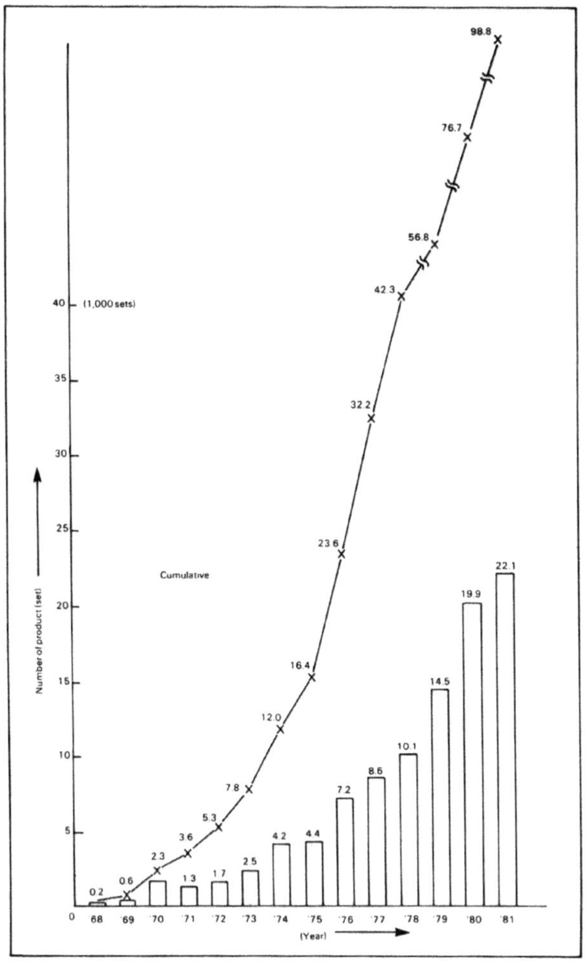

Fig. 1. Production of robots by unit.

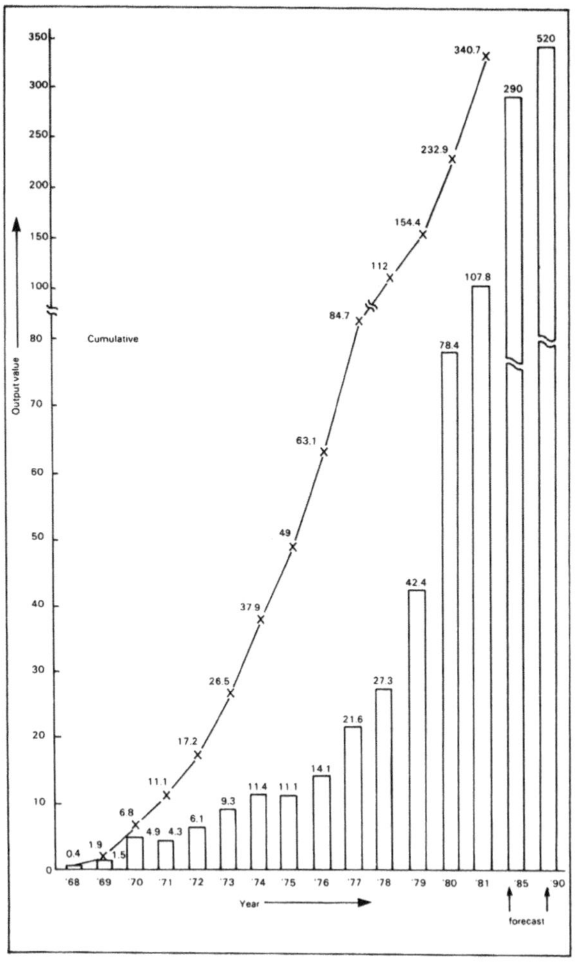

Fig. 2. Production of robots by value (x 10^9 Yen).

businesses to invest more in automation.

☐ Third, Japan's population has showed a 1.1% increase per annum, whereas the growth rate of its labour force has levelled off at 0.7% chiefly due to the mounting number of students going into higher education. This has increased the need for higher productivity to achieve the 5% annual economic growth target. The need is particularly felt in the manufacturing sector where, unlike the tertiary industry, an increase in labour force is not expected (Table 2).

☐ Fourth, shortages of skilled labour have again reached considerable proportions due to a growing number of young people moving on to a school of higher grade. In June 1980, skilled labour shortages amount to about 840,000.

☐ Fifth, there is the need to prevent industrial accidents and occupational diseases due to dangerous work as well as unfavourable working conditions. To meet this need, great pressure has been placed on management to increase the use of industrial robots.

Finally, organised labour in Japan is not primarily in the craft unions but in industrial unions or unions organised on a company basis, making it much easier to introduce industrial robots and relocate workers.

Table 2. Changes in the Labour Force by Industry
(Unit: in 10,000)

Industry	*1975*	*1980*	*1985 (est.)*	*1990 (est.)*
Primary	661	577	505	401
Secondary	1,841	1,926	1,962	1,970
– Manufacturing	1,346 (25.8%)	1,367 (24.7%)	1,377 (24.0%)	1,334 (22.5%)
Tertiary	2,721 (52.%)	3,033 (54.6%)	3,269 (57%)	3,558 (60%)
Total	5,223 (100%)	5,536 (100%)	5,736 (100%)	5,920 (100%)

est. Prime Minister Office "Survey on Work Force"

The major industrial robot users are:

- The electrical machinery manufacturing industry (whose share of total distribution of industrial robots in 1981 was 32%),
- the automobile industry (30%),
- the plastic molding and processing industry (9%),
- the general machinery manufacturing industry (8%),
- the metal working industry (5%).

In addition, the textile, chemical, steelmaking, and ship-building industries are users, indicating that industrial robots are demanded in a wide range of industries in Japan. However, demand for industrial robots in non-manufacturing industries accounted for only 1% of total demand in 1981, while exports stood at a meagre 6% of total output that year.

Industrial robots were applied in 1980 and 1981 as given below:

Process	*1980*	*1981*
plastic moulding	8000	7500
machining	3200	4200
assembly	2800	2900
welding	1600	2600
press	800	1100
casting	700	1000
painting	130	350

It is expected that automation of assembly processes in small-batch production in different sectors of the machinery industry will be possible in the future as the technological develop-

ment of intelligent robots makes progress.

Exports of industrial robots are expected to reach 16% of total output in five years from now and 17% in another five years. For this to be true, however, Japanese makers will have to team up with overseas engineering firms, since industrial robots require highly sophisticated systems engineering prior to installation and also good maintenance services after they are put into operation.

'Wide use of robots expected'

Aside from the manufacturing industry, the wider use of robot technonology is expected in areas such as:

- ☐ Nuclear – for inspection and maintenance of nuclear power plant, and radioactive waste disposal.
- ☐ Medical and social welfare – to help the physically handicapped (1,989,000 in 1980, of which 1,130,000 are crippled) to work as efficiently as the average person, and to assist the bedridden (530,000 in 1979) in making their living.
- ☐ Ocean exploitation – for a variety of handling operations needed in underwater construction, machining and geological surveys, and for multi-purpose observations.
- ☐ Agriculture and forestry – for a number of agricultural and forestry jobs including fruit harvesting, crop-dusting, lumbering, trimming, and lumber collecting.
- ☐ Construction – for assembly of reinforcing bars, painting of bridges, and interior and exterior finish work in high-rise buildings.
- ☐ Transportation and other service sectors.

In order to strengthen international competitiveness and profitability, in the years to come productivity improvement is an essential consideration that businesses should take into account. Particularly in Japan, which is lacking in natural resources and is thus supporting its national economy by means of processing trade, there is a pressing need to improve productivity.

Also, there is a need to prevent industrial accidents and occupational diseases and to upgrade the quality of human labour. Mankind is turning into a slave of machinery as exemplified in the conveyor system and there is a need to reverse this degredation of humanity. The provision of agreeable working environments and the humanisation of working life are an important key to promoting industrial and social welfare.

'Robots upgrade product quality'

Industrial robots provide a powerful means of attaining these goals. The application of industrial robots has a wide variety of economic and social impacts. Characteristics such as flexible movement functions enable small batch production to be automated and are causing a change in the production system from man-machine to man-robot-machine. These changes are improving the working conditions of human labour.

Industrial robots not only improve productivity and welfare but they upgrade product quality, save resources and energy based on more effective capital investments and a reduced rejection rate, improve production control, stability of employment, solution of skilled manpower shortages, creation of new technology, development of new industrial sectors, and so forth.

He who dares in technology, wins

Professor Wilf Heginbotham, director general of PERA, looks at the way robot research has to go

ARE INDUSTRIAL ROBOTS the engineering paradox of the age, the panacea for present industrial ills or are they the prophets of future development and the heralds of a new free and enlightened society?

Robots are paradoxical because in their present form they are pale substitutes for their human counterparts; and they have a very low level of cybernetic capability. They cannot cope with significant random variabilities in the workplace and are unable to solve complex industrial problems where decision-taking forms an important part of the activity.

The panacea can be recognised from the fact that these devices concentrate the mind greatly because of their inherent stupidity. They force engineers, managers and the workforce to recognise that 'order out of chaos' is an essential prerequisite to success.

It is interesting to note that prophesies can be extrapolated from the past as follows: The automation (or mechanisation) of yesteryear depended very much on the ability to fragment industrial operations into small steps, and by combining groups of these small steps entities which could be understood by the unaided intellect of individual human beings were possible.

Human limitations were therefore the reason why the fragmentation and breaking down of production functions was necessary in the first place. This process is now in retreat in no uncertain manner. Using intellect *no greater* than that required to deal with the fragmented production methods of the past plus the data processing power of the computer, allows more complex entities of the production function to be encompassed, dealt with and controlled by the same workforce using relatively simple alpha numeric communication.

These prophesies can be verified by the progression of events from the past in the context of the first Industrial Revolution which dictate that the distribution of skills will have to change.

For instance, the manipulative skill required to use a traditional spinning wheel was replaced by the more *numerate* skill required to operate the spinning jenny, the mule, and latterly the ring frame. This numeracy was needed to calculate the change wheels required to control the spinning of thread to put in the required amount of stretch and twist per unit length of the thread by mechanical means.

Certain levels of interaction were necessary also because operations had to take place at that time within the uncontrolled environment of the early cotton mills, changes in temperature and humidity required the 'minder' to alter the change wheel ratios even when no other changes were evident in the basic raw cotton. This was to avoid uncontrolled thread breakages because of the different behaviour of the thread fibres under different conditions.

Professor Wilf Heginbotham.

Changes of 'staple length' between different batches of cotton also necessitated different change wheel ratios for successful spinning and a new semi-instinctive 'feel' (or skill) based on experience had to develop.

The situation is also akin to the concept of 'the toolmaker' who manufactured the first fixtures and special tools to replace the direct skill of a human operator; that is the new skill was transferred to an elitist few. The NC machine tool carried out this skill transference a stage further by reducing the need for 'hard' tooling and transferring the skill to the 'part programmer'.

This procedure has now moved full circle because the machine operator/setter can do all this by transferring his *machine manipulating* skills directly to the machine through a microprocessor link. Thus a new skill has now been developed by the system designer who is really a kind of software 'toolmaker'.

The current microchip revolution is therefore only an extension of these trends. However, the level of numeracy possible is many thousand times more than that required by the spinning machine minder, and the software reduces the complex industrial situation to the level of a 'computer game'. Such possibilities can replace the 'feel and experience' of human operators and it is in this capability that the industrial scene will undergo a dramatic change.

To be able to achieve effective integration of smaller batch production however requires the ability for 'machines to talk to machines' without human interfacing and even with present day first generation robots this is possible. The situation is somewhat similar to the early days of numerical control, where the control systems for different machines were incompatible and until this problem is solved for robot technology the prophetic implication of the robot as an integrating influence will not be attained.

Many of the present problems of factory integration are concerned with uncontrolled or uncontrollable variabilities – spatial, product state or time and these are generally most intractable at the workplace.

Progress will be made therefore only from this point in time by developments to make the mechanical functioning of the workplace capable of adapting itself to deal with the actual variations concerned directly with the doing of real work. The development of cleverer and cleverer vision and tactile systems does not help in this respect because the computer working space is vastly more versatile and capable of adapting to change than the mechanical interface at the workplace.

Like any other industrial development success will only come from a cascade of simple cost effective applications. Progress will not be made by dramatic leaps in technology. Robots and computers will only obey our misguided whims and will not correct for them. Future development is vitally dependent on industrial courage and a willingness to participate in this evolutionary process – 'he who dares wins'.

When it will be important not to lose anything

Stanley Polcyn, senior vice president of Unimation, looks at market prospects and new developments in the US

THE revolutionary gains robots have achieved in productivity and product quality have opened wide the eyes of industrial managers, but they are only a glimmer of their potential for restructuring the workplace.

As managers have grown enamoured with what automation can do for them, robots are poised to move into a vast array of jobs that a few years ago seemed ill-suited for their limited capabilities. But the pace of technological development in robotics has quickened, leading to sophisticated controls that will expand the role of the robot as industry's partner in productivity.

A number of driving forces are converging to produce the much heralded 'explosion in robotics'. First, the decreased cost and increased dependability of robots is making them more attractive economically. Worldwide competition among robot manufacturers is softening prices. Low-cost computer controls have been added to many systems and sensor-controlled robots are preparing to leave research and development departments. Most importantly, management and workers are beginning to realise that the robot will improve the quality of work life.

'The population will explode to 100,000'

All of this tends to fuel the optimism that surrounds projections of robot growth, which is expected to be 35% annually. According to some projections, the current US robot population of 6,000 will explode to 100,000 by 1990. That means the $155-million a year market could mushroom to a multibillion dollar annual market in the next decade.

Long standing robotic applications in spot welding, machine loading and fusion welding will continue to anchor US and European markets. In those countries with significant industrial bases, robots will create employment by opening new opportunities for workers engaged in related aspects of automation such as warehousing.

U.S. ROBOTIC APPLICATIONS*
December 1982

Spot Welding	1,190
Arc Welding	270
Painting	290
Finishing (polishing, grinding)	30
Assembly	50
Machine load & unload	1,470
Die Casting	880
Investment Casting	120
Material Handling	1,950
TOTAL	6,250

*Above information compiled by Robot Institute of America

Although the application of robotics to metal working industries will remain strong, new markets will emerge in a wide range of operations. Small parts assembly, sophisticated material handling, adhesives dispensing and electronics are among the fields where robots will have a significant impact.

Hospitals and service industries will benefit from robotic technology. Robots have the capability, for example, to work at a patient's bedside to provide service during certain situations. In other areas of the hospital, they can provide technical assistance where they have the ability to work in high degrees of accuracy and where such precision is required as in surgery.

Robots will replace some workers, but they will spearhead a new drive for cost reduction efficiency among those countries with large industrial bases who have lost export markets. In doing that, robots will permit segments of the industrial base to grow.

The US, for example, has had an unfavourable imbalance of trade with Japan of $12 billion. That represents 40,000 jobs per billion dollars or 480,000 jobs that the US has lost to Japan.

Robotic technology can lead us back towards the recovery of jobs that would otherwise be lost to US workers. One way this will be achieved is by stimulating growth in the manufacture of hardware and software that will generate new jobs and expand the industrial base of the US.

Among the markets undergoing considerable growth in robotics is the glass industry. It will feel a sharp impact from the use of robots in a wide range of applications, including moulding, straightening, transferring, palletising and packaging. Glass manufacturers are increasingly relying on the flexibility of robotic technology for short and medium sized production runs. For example, robots can be applied to packing operations that require transferring of glass jars for filling of products.

In one of the first applications in the glass industry, a robot has been called upon to use dexterity, delicacy and precision in England. At the glassworks of Chance Brothers, a member of the Pilkington Group, a robot lifts molten glass from a high temperature furnace on the end of an eight-foot gathering arm in producing radar equipment.

At a Ford Motor Company plant in the US, another robot is an integral part of the assembly line where it transfers automobile windshields from a laminating process to a carrier. This carrier takes it to another station where it is installed in the frame of a car.

'Using robots to lay loomboards'

The growing potential for applying robotics to assembly operations has been demonstrated at a British manufacturer of forklift trucks. Lansing Bagnall, Limited, is using a robot to lay wire on loomboards as part of producing electrical wiring for these trucks. The robot performs a variety of repetitive tasks to insert pins and then lay the wire on the loomboard.

Similar versatility is required at a California electronics firm which is the inventor and leading supplier of floppy disk drives. At Shugart Associates, an industrial robot performs eight steps in assembling the mechanical core of the firm's floppy disk drives and comparable products. The robot builds one assembly in 78 seconds compared to the two minutes that is required by three persons.

Continued development of sophisticated controls will power the new crop of robots emerging from research and development laboratories. These controls will enable them to take on more jobs on the factory floor than those their predecessors were capable of handling. New vision systems will equip robots to move into more complex operations. Unimate PUMA robots demonstrated at last year's Robots VI Conference and Exposition showed how they could be interfaced with Univision I for assembly, bonding, inspection and palletising operations.

Considered a major breakthrough in automatic arc welding, the Unimation Univision II system allows arc welding on parts that were previously impractical to weld automatically because of poor fit or expensive fixturing. This two-pass seam tracking system uses a fibre optic sensor to 'look' at the seam to be welded and automatically adjusts the robot's actual weld path. The costs must still be brought down and the technology has to be enhanced, but we are within a few years of achieving these things.

Three dimensional vision exists today but it is far from reliable. When it is perfected, however, it will enable robots to overcome a limitation that has prevented the first vision systems from differentiating parts that touch. When coupled with tactile sensing, opportunities for robots will expand in operations such as bin-picking, assembly and disassembly.

'Mobility helps robot serve more areas'

Although research in robotics has emphasised vision, such force feedback development is essential for robots to take their place on still another level within the factory. In research labs, grippers have been designed to pick up such objects as an egg without crushing it. The ability to stop robot motion after reaching a force threshold has been shown. By gaining better control over the forceps at the gripper level, we will be able to program robots to load and unload areas within a warehouse and package items for shipment.

By the end of the decade, mobility will enable robots to serve more areas of a factory and perform tasks in the household. Machine loading and servicing, clean-room material handling and maintenance are areas where mobile robots could improve productivity in a plant.

The limitation to mobility is that robots require a highly structured workplace.
The work environment will have to be arranged so that objects are not misplaced. If they are, they could confuse the sensory-based microprocessors interpreting the data.

But the control of such factors lies within our grasp. And so does the automated factory. The components of the automated factory exist in the form of sophisticated cutting tools, robots, computers and material handling systems. They need to be joined as integral parts in an automated network. We need industrial managers with the vision to see the potential for linking these islands of automation that have emerged. When they are joined, industry will have a powerful partner in productivity.

The European market: Still looking good

The robot associations in the United Kingdom, France, Sweden and Italy all report growth markets.

THE British Robot Association announced in February that the UK robot population stood at 1,152 by the end of 1982. This represents a growth of 439 during the year.

The figures are based on returns from all robot suppliers in the UK and are claimed to be one of the most accurate set of statistics on robots produced by any country. Overall the growth in 1982 compares with 342 in 1981 and is an increase of 61%.

The striking feature about an applications breakdown for the UK, as shown in Table 1, is the even spread, albeit a rather thin one. Spot welding takes prime position but arc welding, injection moulding and machine tools have had similar increases. The numbers in spot welding are in maybe only two or three installations whereas the others are spread over many more companies. Spot welding represents 22% of the UK robot population.

Spot welding robots, at present are almost entirely employed in the automotive industry, particularly for car body assembly. The UK's auto industry has been undergoing dramatic changes in recent years which has meant a lower output compared with some of its competitors. It is only in the last two to three years that spot welding robots have been installed. This state is reflected in the car production and robot figures for 1981 shown in Table 2. These data highlight just how well ahead the Japanese are in the use of this form of automation. Of the four countries covered in this survey Japan has 39% of the car output but an enormous 62% of the robot spot welder population.

Perhaps the one bright spot in the UK figures is for machine tool servicing. During 1982 it increased by nearly 100% to 111, maybe indicating a more active interest in FMS in the UK.

There is now a significant number of real robots in research and education which bodes well for the future. Graduates with direct experience of the 'real thing' cannot do anything but good. There was also a rapid increase in the 'non-industrial' teaching or hobbyist robot which many companies and educational institutes are buying as a first step into robotics. In 1982 the total stood at 762 and it is likely that their number will exceed the industrial robot population by the end of 1983.

Table 1. Robot Applications in the UK. (source: British Robot Association)

	total at Dec. 1982	installations during 1982
Surface Coating	124	36
Spot welding	249	82
Arc welding	157	70
Grinding/Fettling, etc	13	7
Assembly	32	17
Die casting	36	3
Injection Moulding	167	78
Machine Tool Servicing	111	55
Press Tool Servicing	27	3
Inspection/Test	15	8
Palletising/Packaging	38	14
Education/Research	38	28
Investment Casting	12	4
Forging	5	3
Other Handling	10	7
Other	118	24
Totals	1,152	439 (+61½%)

The country of origin of the products in use in the UK is often a source of discussion these days. Robots are no exception to this and the figures are no less alarming. The number of foreign built machines now amounts to 75% with other European countries being the major suppliers. It appears that USA-sourced machines are proportionately declining in favour of Japanese.

Although the latter stands at only 14% of the total machines installed it did take 25% of the 1982 market. It is likely to continue increasing through the many deals now being established between UK and Japanese companies. There are 12 company links between Japan and UK, as shown in Table 3, with additions occurring apparently every month. As this includes some recent major agreements, notably involving companies such as GEC, 600 Group and Evershed the UK 'Japanese-robot' population is bound to grow.

This type of growth is not alarming the British Robot Association since its main aim is to encourage the use of the technology by industry. Whilst it would like to encourage the manufacture of robots in the UK it would much prefer to see a foreign robot rather than no robot at all. The Association only hopes that the big AUTOMAN showcase at the NEC in May will encourage more of British industry to make better use of the availability of robots.

Table 2. Car and robot figures for 1981

Country	Passenger car production*	Number of spot welding robots
UK	954,650 (5.4%)	167 (3.4%)
W. Germany	3,577,807 (20%)	771 (16%)
USA	6,253,138 (35%)	900 (18%)
Japan	6,974,131 (39%)	3,055 (62%)

* Source: SMMT.

The UK Government through its Department of Industry is giving increasing support to robotics in the UK. So far over £8,000,000 has been given in aid to industry. Also seminars have been organised on a regional basis to introduce local firms to the subject and to explain what support can be obtained. There have been 24 so far averaging 50 participants.

The largest slice of the D of I 'cake' has gone to applications. A total of 75 projects involving 143 robots have been approved at a cost of £5,500,000 to the UK taxpayer. There has been £2,400,000 allocated towards 20 robot manufacturing projects, which includes £1,500,000 to Unimation to develop a new range of robots as part of its expansion at the Telford factory.

To this must be added £104,000 given for consultancies to which D of I

Table 3. UK-Japan Link Ups (source: British Robot Association)

UK company	Japan company	Type of arrangement
Airstead Industrial Systems	Hirata Industrial Machineries Co.	Sales (UK)
	Daini Seikosha	Sales (UK)
Norman Butters	Osaka Transformer Co.	Sales (UK)
Dainichi Sykes Robotics	Dainichi Kiko	Sales (Europe) & manufacture
Evershed Robotics	Toshiba Seiki	Sales (Europe)
GEC Electrical Projects	Hitachi (process robot)	Sales (Europe), manufacture & technical
Haden Drysys International	Hitachi (painting robot)	Sales (Europe)
Harmo Robots	Harmo Co.	Sales
Haynes and Fordham	Star Seiki	Sales (UK)
	Shin Meiwa Industry	Sales (UK)
IBM	Sankyo Seiki Manufacturing	Sales (UK)
Lansing Industrial Robots	Hitachi (process & 'Mr Aros' robots)	Sales (UK)
Lincoln Electric	Yaskawa Electric Manufacturing Co.	Sales (UK)
Ringway Power Systems	Mitsubishi Heavy Industries (through Voest Alpina, Austria)	Sales (UK)
600 Fanuc Robotics	Fanuc	Joint venture – sales (UK) & manufacture

provides half the cost of a study lasting up to 10 days. These can entail feasibility studies, evaluations, design exercises or other projects by D of I approved consultants.

Neither is robotics research in the UK being neglected. In 1981 the Science and Engineering Research Council started a robots initiative programme and now has over 20 major projects running. These are all joint ventures between a university, polytechnic or a technical college and an industrial company. The S.E.R.C. funds to the research institute are supported by a significant contribution, in some cases up to 70% of the S.E.R.C. support, from the company. This aid from the industrial partner may be in the form of cash, loan of equipment or manpower.

This degree of support and activity points to a good future for robotics in the UK. But it is considered essential by the B.R.A. and the Government if the country is to match up to the manufacturing problems of the 1980s. It is seen by many that robotics and similar technologies are the only way that British industry can compete with Japan and the other emerging nations of the Pacific.

West Germany

IN spite of the present difficult economic situation there is no recession to be seen in the field of flexible automation like industrial robots. Indeed, in spite of the economic situation – or perhaps because of it – the number of industrial robots being utilised in the Federal Republic of Germany is increasing fast. The number of installed industrial robots jumped by 1,200 for a total of 3,500 in December 1982.

The programmable industrial robots – first discovered around 20 years ago, have been utilised for 10 years now, many in the field of production technology. The programmable mechanical units are establishing themselves also slowly in smaller and medium size business applications. Even with the seemingly necessary large investments industrial robots have become an important part of our industrial production processes and should not be forgotten as a feasible automation possibility.

The industrial robots are recognised today almost worldwide as universally adaptable, multi-axis automation devices which can be expanded to many movement positions with use of one or more arms or joined to other machines to create a multi-machine assembly system. Their movements must be programmable to execute difficult movement processes such as angles without mechanical encroachment in the control process. They can also be programmed to use sensors.

The industrial robots can normally be equipped with grippers, tools, measuring devices or other production devices and are capable of performing handling and other production processes. This definition is important to note because unprogrammable pick and place units have been mixed in some cases with programmable handling devices. This has led to confusion in analysing the present-day industrial robot market.

The effort to make the working world a better place, improved devices, higher wages and intensified research efforts have all led to expansion in the use of industrial robots and will continue to lead to further expansion.

Until a few years ago the number of functioning industrial robots was very small because the demands of the different production fields could be only partly met. Vastly improved devices, much improved and inexpensive control technologies and increased reliability have enabled industrial robots to become *flexible means of automation* which represent a good and expedient alternative to one purpose automation.

The flexibility of automation, the possibility of economical application and the concern for the quality of the working environment are the most important reasons why industrial robots are being used more and more. The economic and social considerations are strengthened when we consider capital intensive installations which must function several shifts every day.

According to reputable studies of the working environment it can be very harmful to the well-being of an operator to work on the third shift. Automation with handling devices can make it possible to carry out the third shift without human help.

Because the present-day industrial robots are not all capable of operating without human supervision efforts are being undertaken throughout the world in order to solve this problem. The flexibility of the necessary periphery, ordering and loading arrangements must be made compatible to the industrial robots.

We must design, for example, ordering and programming set-ups which produce the required output for an entire shift. There must be a supervision system which watches over the whole assembly and shuts off or implements an alternative programme in the event of a breakdown.

The industrial robots, although originally conceived for small and middle sized tasks, have been utilised mostly for mass production. The possibilities for industrial robots to perform spot welding and coating tasks have been only partly realised although industrial robots are really the correct and most efficient solution here. They can be programmed to process large numbers of identical units clearly and simply and their changeability can lead to significant time and cost savings.

The industrial robot makes it possible to carry out on-the-spot special constructions in a quick and affordable way. The robot can be used also once substantial changes have been made and which, without flexible industrial robots, would otherwise lead to useless, outdated and expensive machinery.

Although 30% of all industrial robots are presently being used in the automobile industry it should not be assumed that the financial power of a large corporation is necessary to afford such a robot. This high percentage is due mostly to the large number of spot welding robots being utilised in the welding production lines of the auto makers.

In other fields such as coating, arc welding, loading and unloading of machine tools or from other machines and testing units there has been a sharp increase in the use of robots, especially in medium sized establishments. According to a study in 1980 every ninth industrial robot was being utilised in a business with less than 500 employees. Although there are no newer studies it is safe to say that today every seventh or eighth robot is being utilised in a business with less than 500 employees.

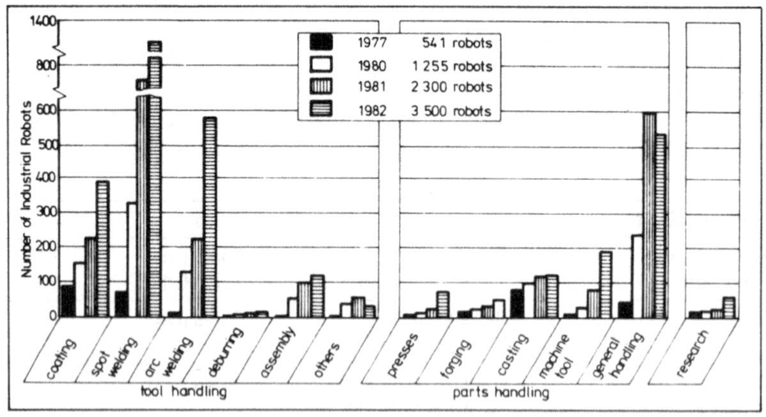

Fig. 1. The growth of robots in West Germany from 1977-82.

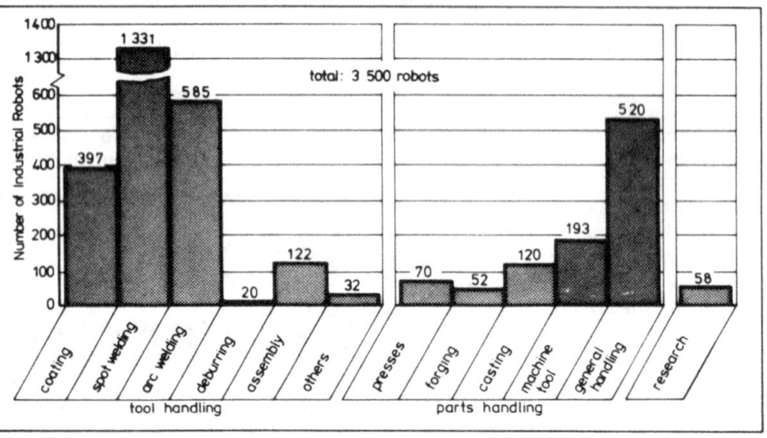

Fig. 2. Number of robots installed for each activity.

The division of the application areas is usually into the two fields of machine tool handling and workpiece handling. In machine tool handling the industrial robot executes, for example, a spot welding unit or a spray gun. In workpiece handling the industrial robot is equipped with a gripper which either sets or removes workpieces in a machine.

Fig. 1 and 2 show the change in application fields in recent years and the number of industrial robots presently in use in the Federal Republic of Germany. It is worth noting that the 3,500 industrial robots shown for December 1982 are installed units. The number of sold but not yet installed units is clearly more.

The strong increase of spot welding units is due especially to the use of new body welding production lines in the automobile industry. In the field of coating the steady increase is due to the fact that industrial robots are now considered a necessary part in coating technology.

The large expansion in continuous path welding is due to improvements, especially in continuous path control which have also made it possible to utilise industrial robots for arc welding.

In spite of intensive devèlopments at

IPA the utilisation of industrial robots for deburring and casting fettling work has not yet been perfected in spite of the urgent necessity in this field. The sensors must still be tested under normal working conditions before an improvement will be seen here.

The increase of assembly robots is due to the significance of assembly activities in West Germany. In this case the industrial robots are simply not capable of performing certain tasks due to lack of sensory capabilities. Therefore there are relatively few industrial robots capable of performing assembly tasks.

In work piece handling there has been an increase in almost all areas. Here it should be noted that beside the loading and unloading of machine tools robots are included in the loading and unloading of measuring and testing machines as well as pallet and commission work. The special first-time recognition of this field of loading and unloading shows that industrial robots are clearly capable here in comparison to the devices which have been utilised up to now.

The large number of industrial robots in the field of research and testing shows clearly that the necessity for research and development in this area is being taken seriously and will be taken further. These industrial robots can mostly be found at the manufacturing firms and at the larger industrial robot utilisers but they can also be found in research and development laboratories. For example, at IPA there are some 15 industrial robots.

The further expansion of the industrial robots will mainly depend on two developments. Firstly, we must investigate further possibilities in fields in which industrial robots have been up to now scarcely or not at all utilised. This must be done in smaller and medium sized establishments where hindrances of technical and economical nature must still be overcome. Secondly, we must develop still further the systems which are already functioning. Necessary further developments here include improvement of reliability of the industrial robots and increased periphery and sensory technology.

The further development of sensor technology and adaptive robot controls to process sensor signals will open up new fields as well as increase the flexibility of present ones. When one considers the social considerations, such as the possibility to improve the working environment, it can safely be said that industrial robots will be used in production areas in the 1980's as NC-Machine tools are used today.

The industrial robot is totally qualified to become an important export item for the German machine industry. This is, in fact, already coming true; some German robot manufacturers already show an export of 30–50% of their production.

However, the industrial robot cannot and should not be permitted to become an all-healing solution for all automation problems of the future, but it has strengthened its application field in production and is, through its flexibility, in a position to strengthen the competitiveness of our industry.

France

FRENCH people in general, and French technicians in particular fancy they have the best of almost everything: the best culture, the best women, the best wines, the best trains, the best aeroplanes, the best guided missiles, the best television, and of course the best robots.

Without looking into the other claims, it is worth examining the state of robotics in France. If we avoid retracing our steps back to previous centuries (although the marvels made by Vaucanson, Maillardet and Jacquard were robots of the highest class since they were re-programmable and sometimes had continuous path control) we find that French robotics began with the very strong nuclear effort and the telemanipulators developed for this field during the 1960s and early 1970s.

Since then the robotics group of the French Atomic Energy Commission (CEA) has produced a number of highly sophisticated prototypes for teleoperation (nuclear, underwater and medical applications) and for automated inspection (such as Virgule, Merit and MAM) which are well known in international research circles but which do not pretend to be fast-selling pieces of equipment.

In 1974 the Ministry of Industry started the first French national research project in robotics called Spartacus. This was followed in 1980 by the second one – ARA or Automatique et Robotique Avancee) – which should last until 1985.

These efforts, now running at close to FFr100 million a year in Government spending – have placed the country in one of the best positions in terms of research potential in robotics. The schemes involve some 300 researchers who are financed by public funds.

However, the first results of this work at the industrial level are only just starting to be seen with the transfer of products and people from research to robot manufacture, and with the creation of high technology by researchers.

In parallel with the research effort and at about the same time, that is in 1974, another effort was started at the industrial level with the development of the Renault range of industrial robots. In less than three years, high performance hydraulic robots with the most modern computer controls were developed and put on to spot welding lines.

Since then, research and development has been going on with heavy financing from Renault to produce some of the most advanced robotic techniques now in use in factories. However, it will be many years, if ever, before these investments will be paid back by the ACMA Cribier and Cybotech subsidiaries which market and sell the robots.

Soon after Renault began its effort a number of other companies started to produce industrial robots with most of them being aimed at the very high end of the technology – and hence the market. Although more than 30 companies are now selling robots the most famous French robot companies are AKR for sophisticated painting robots, Sciaky and ACMA Cribier for spot and arc welding machines, and Sormel and Scemi for assembly. The last one is already famous for its LM language developed at the University of Grenoble

– the only one of its level available on the market.

However, although some of the companies are now very well known on an international basis the following chart indicates that while their contribution to the world robot population is not yet over-whelming in a short space of time it could be not insignificant.

Robot population in France numbers between 985 and 45,000 depending on definition. It is not worth arguing on this topic at this stage since here we are concerned only with the 'highest' class of robots: that is those that are servo-controlled and computer driven. This does not mean that other robots are less interesting (though they are for French technicians) but simply that accurate statistics are available only for this class of robot.

(I challenge any other country to prove that its statistics for these other classes of robot are accurate.)

Indeed, it may be possible still to prove that the largest market (in value) is for the simplest form of robots – that is, providing you agree to call them robots.

If now we turn to the areas of application of sophisticated robots in France it is not difficult to guess that they are to be found mostly in the automobile industry. However, the picture is changing rapidly because of two important factors:

- ☐ The introduction of robots into small firms
- ☐ The arrival of assembly robots

The first factor leads to a large potential market for small robots where the constraints of flexibility, productivity, and quality combined with the increasing cost of labour, now bring a definite advantage to the use of robots in many applications, such as handling, continuous welding, spraying and so on, but only as long as the investment is not too large.

The second factor is mainly of interest for large companies involved with assembly. Already the first applications are now being experimented with in France and they concern the electro-mechanical industry (electric motors, relays, appliances and so on); the electronics industry (computers, telephones and home electronics); and of course the automobile industry (various components such as gearboxes, engines, dashboards, brakes and so on).

Another factor which may change the distribution of robots in French industry in the more distant future is likely to be the opening of new fields of application.

Although now at the experimental stage in France it is nevertheless

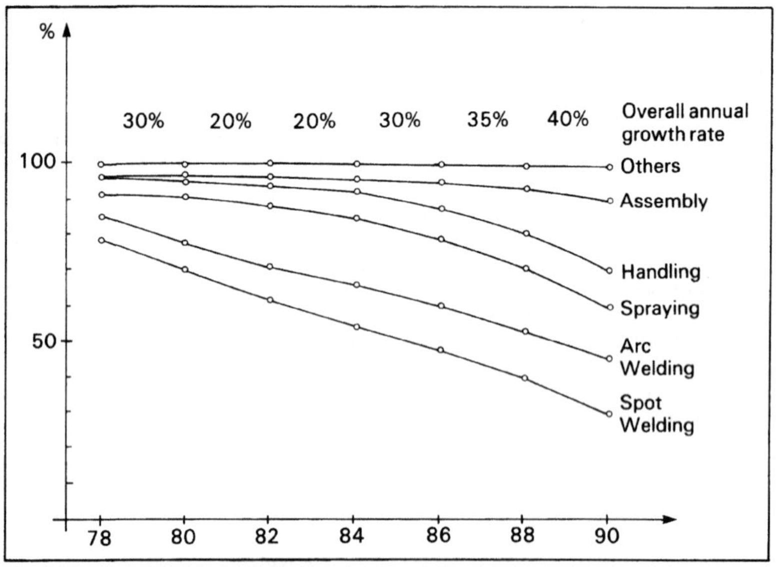

Fig. 1. The changing pattern of robot market share in France.

Table 1: Actual and Estimated Production of Computer-Driven Robots

	Production		Expected Production		
	before 1982	1982	1983	1984	1985
ACMA	260	120	160	300	400
AFMA	0	9	40	80	100
AKR	17	28	80	160	350
SCEMI	0	4	50	200	300
SCIAKY	7	20	20	20	30
SORMEL	8	10	20	50	100
TOTAL	292	191	370	810	1280

Table 2: French Robot Population of four types of robot (1982 estimates)

	A	B	C	D
French	1500/2000	10000/15000	550	30
Foreign	2000/2500	20000/25000	400	5
Total	3500/4500	30000/40000	950	35

A = telemanipulators
B = pick and place – fixed and variable sequence
C = programmable – servo controlled
D = intelligent robot

possible to point to a number of areas of application for robots:

- □ The farming industry (food harvesting, selection and preparation)
- □ The meat industry (butchery)
- □ The fishing industry (sorting, preparation and packing)
- □ The textile industry (handling and three dimensional sewing)
- □ The wood industry (cutting, sanding and assembly).

However, it is far too early to predict any figure on the introduction of robots in these fields, even if the technology is available. Also, of course, the socio-economic factors of introducing robots into these industries need to be taken into account, especially where labour is of national importance.

A French Government project called AMES (Automisation et Mutation Economiques et Sociales) has now just begun to evaluate these factors and to propose some guidelines at the national and international levels.

At the international level who will benefit most from these expanding markets? When raising the subject in Europe it is instinctive to look towards the US and in the other direction towards Japan. Added to which it is relevant to be concerned with the international strategies of the multinational companies groups such as IBM, General Electric or Westinghouse.

However, the problem is more complex than just the control of the hardware and software markets. In order to introduce robots on a large scale in any country it is necessary:

- □ To have a strong and healthy industry ready to accept and invest in automation
- □ To have engineering firms which can look into production problems in their globality and propose solutions
- □ To have component manufacturers which can cover the complete scope of automation technologies
- □ To have excellent maintenance service for all these components (a manufacturer cannot afford to lose several days of production waiting for a spare part).

The robot is of course only one component in the field of automation. Besides which the robot comes in many different forms, each one best suited for certain applications.

No one firm and no single country (not even France) can pretend to manufacture all the best robots and all the components and techniques to meet every need. Therefore exchanges of product, technology and services will be necessary.

In particular it will always be imperative to have consultancy and maintenance on a widely distributed basis. So, through healthy and voluntary collaboration between companies and countries it should be possible to share out the work still needing to be done in this exciting field.

So, if you are looking for some of the highest technology in robotics and production automation then come and see us in France...

Sweden

SWEDEN is one of Europe's leading users and manufacturers of industrial robots. On a per capita basis it is as good if not higher than that of Japan. It now has its own body, the Swedish Industrial Robot Association (SWIRA) to promote the interests of robotics in Sweden. This is administered by the Swedish Mekanforbund, the Association of the Mechanical, Electrical and Electronic, Metalworking and Automobile Industries.

SWIRA estimates that the number of robots in use at December 1982 was 1,300. The largest number (51%) are performing machine loading and unloading operations. The other major areas are welding (22%) and painting and finishing (15%).

There are three major robot suppliers in Sweden – ASEA, Atlas Copco and Satt-Kaufeldt. ASEA is a robot multinational with manufacturing plants in the USA and Spain, and is shortly to have one in France. Consequently a high proportion of its output (90%) is either exported or built abroad.

Satt-Kaufeldt is the longest established robot builder in Sweden and one of the first anywhere having produced its first robot in 1968. Since then it has installed several hundred in a number of countries including the UK, USA, Soviet Union and West Germany.

The current investment for robotics is good despite a general decrease in investments. The techniques have been well accepted by management and trade unions alike. Now much effort is going towards developing and building FMS. SWIRA anticipates that use of robots will grow at the rate of 20-25% in the next few years.

For the future more sophistication will be incorporated into robots. The major emphasis is anticipated to be on assembly and test for which small and flexible robots will be developed.

Another Scandinavian country to form its own robot association is Denmark. The Danish Robot Association was formed in February 1982 during a meeting held at the Technological Institute. Its aim is to develop close connection with similar organisations abroad and so exchange technological and similar operational experiences.

The Association means to ensure that Danish industry is not left behind by international competition. DRA emphasises that it is of the utmost importance that firms in Denmark are equal to those in other countries in the implementation and use of new technologies, including robots.

Current installations in Denmark are not large with a minimum of 52 robots working in 20 companies in early 1982. A major fact is the absence of any vehicle manufacture in Denmark so that spot welding, which accounts for 20–40% of applications in most industrial countries, is not represented.

The most popular applications are machine loading, with 23 robots, and surface coating with 10 machines in use. However, DRA admits that these figures should be treated as the minimum since some units could have escaped the survey.

Arc welding is seen to be one of the fastest growing application areas. Five robots were installed for this purpose in 1981 and the first quarter of 1982 compared to only one previously.

So far there is only one robot manufacturer in Denmark. So another aim of the Association is to encourage Danish companies to develop robot equipment as a future activity.

Belgium is another of Europe's small countries experiencing a robot boom. It has one of the fastest growing markets, averaging an increase of over 200% since 1977 when a mere five robots were in use. At the end of 1981 this number had grown to 246 mainly due to some multi robot installations in the automotive industry, particularly Ford's Ghenk plant.

Belgium's robotic interests are handled by the group BIRA Robotics which produced the above figures. It claims Belgium is number five in the world league table if robots per production worker is used as the norm. There are 2.3 robots per 10,000 workers in Belgium which lags the figure for Sweden (17), Japan (10.7), West Germany (2.6) and the USA (2.3).

The potential for growth is also large with the Belgian automotive industry producing 3% of the world's output. On this basis the group predicts a robot population of 500–600 by 1985.

There is also activity on the manufacture of robots and systems in Belgium. Last year saw a major licencing agreement between FN-Robotics and the US robot manufacturer Prab Robots. In addition three indigenous companies revealed products in 1981 – Distribel, ES-Eupen and Woit-Otrico.

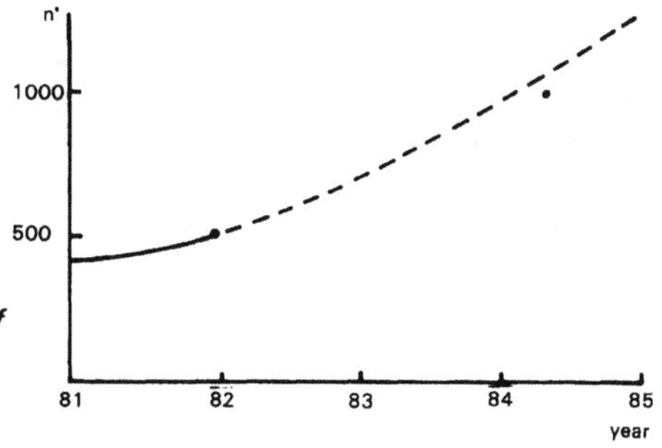

Fig. 1. (Right) Number of robots installed in Italy.

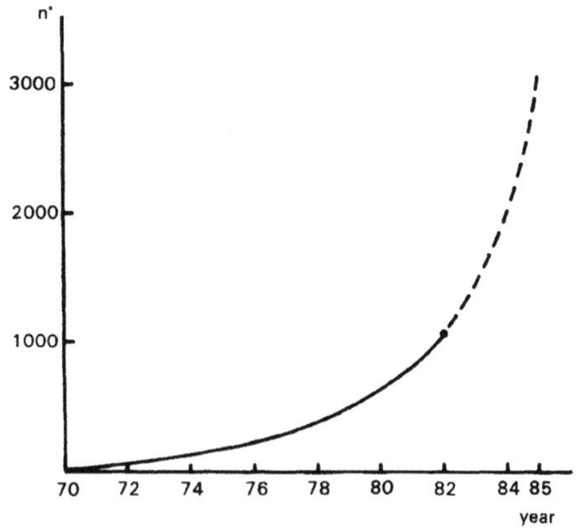

Fig. 2. (Left) Number of robots manufactured in Italy.

Italy

ONE of the most active countries in robotics is Italy, both in terms of use and manufacture. The first machines were installed in the 1960s and now total 1,100 robots. This growth has been aided by Italy's two major motor manufacturers – Fiat and Alfa Romeo. Fiat was the first European car company to install a multi-robot spot welding line.

The widespread use of robots has been helped also by an active Italian robot building industry. The Italian Society for Industrial Robots (SIRI) puts the number as high as 32 at the last count. Amongst these are such acknowledged world leaders as Olivetti, DEA and Comau.

The first robot designed specifically for assembly, the SIGMA, was produced by Olivetti. DEA has produced an outstanding assembly robot the PRAGMA which is now built under licence by GE in the USA. Comau developed the revolutionary Robogate system as well as being the first manufacturer to design a robot for spot welding. Its Polar 6000 robot has an integral transformer mounted on the arm to minimise the problems of trailing cables. Comau's latest robot is the Smart.

Another area in which Italian industry has been more successful at applying robots than most is forging. Again there is a builder who has aimed its product at the specialist problem. Camel Robot has developed two models for this market; this is one reason why there are about 100 installations in the forging area.

The promotion and business of making industry aware of the industrial robot is dealt with by SIRI. This Society was formed in 1975 and aims at promoting the study of scientific, social and educational problems associated with the industrial robot. It is helping to integrate research and development by drawing up a programme for efficient collaboration between research groups and both industrial users and manufacturers.

Currently there are some 30 research centres active in Italy, notably the Universities of Milan, Turin, Genoa, Bologna and Rome. Grants for research have been given by the National Council of Research (CNR) and by the Minister of Public Education through its committees on computer and systems engineering. A special committee has been set up to study the problems and to propose guidelines for a national research project.

Past work at both the Universities of Milan and Turin have gained worldwide renown. At Milan, distributed computer control, robot vision, languages, and the application of AI techniques to robotics have been studied. Turin's studies have been directed at solving part-feeding, including an examination of the bin picking problem.

Table 1. Application analysis of robots in Italy (source: SIRI)

Application area	No. installed at Dec. 1982
Welding (spot+arc)	400
Painting and Finishing	100
Assembly	100
Machine tool loading/unloading	200
Die and investment casting, materials handling	300
Total	1100

The present situation in Italy is such as to favour further rapid growth of robotics. SIRI fully expects both manufacture and use to continue to grow at increasing rates. As shown in Figs. 1 and 2 the number of robots in use in 1985 are anticipated to be 3000 and the number built in Italy to be 1500.

Are there really so many robots in Japan?

John Hartley, writing from Japan, notes that the Japanese have adopted robots as workmates but wonders if their use will continue to expand?

ENGINEERS visiting Japan in search of robots have usually been perplexed. On the one hand, the statistics and filmshows have indicated that there is an enormous number of robots in Japan, and that the Japanese have a tremendous enthusiasm for robots. But on the other hand, visitors have generally been disappointed in the lack of robots they have seen in use in factories.

At the same time, many Japanese production engineers seem to be just as sceptical of robots as do people elsewhere. And they explain that they resist using more robots because they are too slow, too expensive, or too unreliable.

So what is the reality of Japan's enthusiasm for robots? First, it has to be remembered that the Japanese robot population appears to be so large because almost any form of automated handling device is counted as a robot under the Japanese definition of these machines. Secondly, the Japanese media always use these statistics but report on NC or playback robots as if the statistics refer to these types of robots only. Finally, almost any robot with some form of sensor is called an 'intelligent robot'.

'The media reckoned robot fever had set in'

However, for all that, the Japanese did begin installing robots in some numbers in the mid-1970s – mainly in car factories where the aim was to reduce arduous manual work. These installations incorporated no new ideas, though they encouraged many small companies to look into robotics.

By the late 1970s the need for greater flexibility was well understood, and companies like Toyota with a conservative background were realising they must use more robots to increase flexibility. This was because following the energy crisis the growth in car sales had become less predictable, with some models needing to be redesigned much earlier than expected. Not surprisingly the car companies set about installing robots in earnest. And the media reckoned that robot fever had set in.

Fig. 1. Body welding was one of the first applications in Japan.

There is no doubt that in many companies this was the case. For example, one of the Japanese delegates to the Robot Symposium in Tokyo in 1981 said that he had been sent by his company president to find applications for robots in his factories producing grinding wheels. Another production engineer, this time in the auto industry, who had visited the Yamaha factories said: 'The engineers told me that they had been instructed by their president to use robots, but they were having difficulty in finding many good applications in assembly.'

Clearly, there are many companies in Japan whose presidents and employees think that they need robots to compete. And in Japan competition is very severe, whether the products are small portable radios or large construction machines.

But people making robots and responsible for the production engineering in large plants take a more sanguine view. Seiuemon Inaba, president of Fanuc, is one of the leading proponents of robots.

Last year he opened a factory to produce servo motors containing 101 robots and staffed by only 60 people. Yet, commenting on 'robot fever', he said: 'The actual situation is not as it is played up by the media. There are many problems to be solved before robots can be used widely – technical ones, cost, and their effect on employment. We do not expect a rapid increase in sales.

Equally keen to refute the impression that robots are the only answer is Minoru Morita, managing director, production engineering, Matsushita Electric. Believe it or not, one of his reasons is the short model life in consumer electronics, which he says is no more than four months with many products!

'We prefer sequence-controlled arms to robots,' he said, 'because they are more reliable and they are also flexible to some extent. We do use NC controlled devices to insert components in printed circuit boards, and these could be called robots. Their individual downtime is about 0.01%,

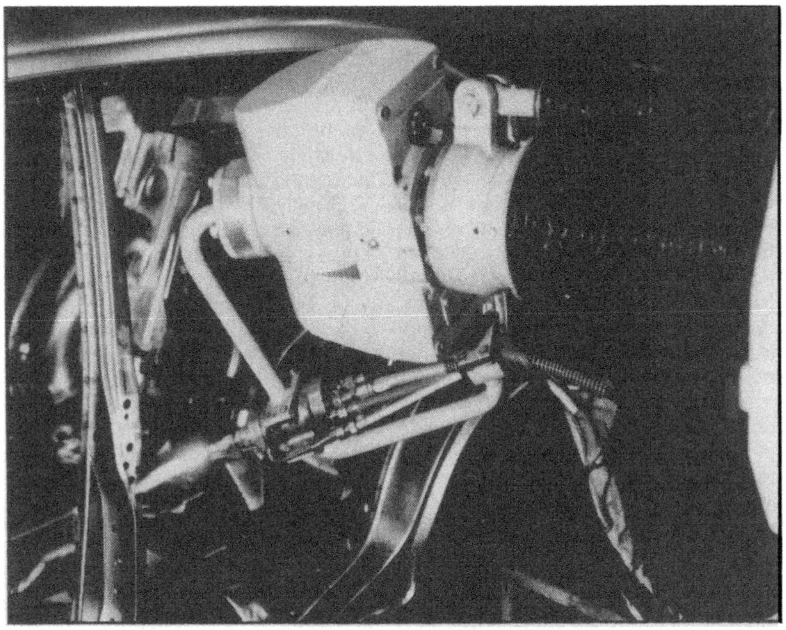

Fig. 2. Gas welding is becoming more common as sensors improve.

so we are able to operate the line with 99.5% uptime.'

Morita doubts whether this would be possible with fully-fledged robots having five or six axes of movement. For this reason he is looking for simpler and cheaper robots to work in his factories and, until these are available, he will concentrate on the use of simple arms.

But it is still a question of horses for courses, with Matsushita developing a whole range of robots for welding, inspection and assembly. Fanuc, Mitsubishi Electric and Hitachi are following a similar path.

Although many people may now be trying to play down Japan's 'robot fever', it is certainly true the Japanese have been much more enthusiastic about the use of robots than Europeans or Americans. The Japanese believe European and American trade unionists will not accept robots in factories, arguing that this explains why there are more robots in Japan than elsewhere. But the reality is different.

When the Japanese car companies installed their first robots they gained a distinct advantage in productivity and quality. And many other people assumed therefore that robots were the road to improved productivity.

Many robots were built and installed to work where productivity gains were marginal. For example, Shin Meiwa and Kobe Steel used their own robots to arc weld without an apparent reduction in manning levels. Daido Steel sold Nissan Shatai a pair of robots for handling at presses where conventional automated hands would have done the job equally well at far less cost.

'Flexibility and high productivity are making robots attractive'

The point is that in Japan the philosophy is to install the robot first, and then make it productive later.

However, we are moving now into a new era where the combination of the need for flexibility and high productivity is making robots more attractive. And because competition in Japan is so fierce it follows that many small companies will install robots simply to keep up.

But large companies are increasing their robot use also in areas where the advantages must be marginal.

For example, in the car industry, the spotlight is now on robot painting and robot assembly. Both Nissan and

Fig. 3. Handling at machine tools is an area for big increases.

Toyo Kogyo have installed robots for spray painting, but in both cases there are doubts about the economics.

For example, Toyo Kogyo has installed 20 Tokico robots in its new paint plant at Hofu. Four of these are used to apply sealer along welding seams, while three pairs of robots are used to apply pvc coating, underseal and anti-chipping compound. These can be justified easily enough because the work is dirty, and the operator is involved in a lot of bending and crouching – in other words, the operators cannot work efficiently for a full shift.

But there are also 10 robots to apply primer/surfacer and colour coats. With such a number of robots, it would be expected that they might apply most of the paint. But not a bit of it; the bulk of the painting is still done by conventional reciprocators, leaving the robots to do the work that cannot be done by normal automation.

In fact, five people are still needed in each colour coat booth to spray inaccessible areas. Outside Japan, it would be difficult to justify the use of robots in such a situation. Management would normally take the view that if men were needed anyway, it would be better not to use the robots.

At Toyo Kogyo employees are aware that the company must improve productivity to match that of Nissan and Toyota, both of which have greater economies of scale. In addition, of course, it was agreed from the outset that the robots would be used to improve quality, and every effort has been made to ensure that his happens.

In fact, the reject rate on painted bodies in the old factory is 10%; while in the new plant it is only 5%, and the target is 2%. Of course, rectification of reject paintwork is very costly, and its reduction can easily justify the robots.

This is a typical attitude in Japan: first, the robots are installed because productivity must be increased; secondly, targets are set to make sure that the robots pay.

Raising productivity is seen in Japan as vital, and managers are often prepared to pay for automation when the actual return on investment seems low. Managers are helped by low interest rates and the fact that robots can be leased at low cost. They do not need to obtain such a high return as European managers to justify the investment.

Fig. 4. Robots are coming into use for spray painting, despite marginal economics.

Equally, there is no doubt that the Japanese are far more susceptible to crazes and trends in manufacturing than elsewhere.

In Europe managers often seem keener to show that a certain piece of equipment will not pay than to find a way of making it pay. In Japan, people will often buy one or two robots with little or no idea of payback.

'Even if robots are uneconomic they will be used'

Even if the robot is not really economic it will be used and investment costs put down to development. And the employees in the shop will be encouraged to come up with ideas to improve the productivity of the robot. If the robot does pay, then more and more robots will be introduced.

Recently, a turning point was reached in robot use at NC machine tools. Until recently few companies were using robots at machine tools, but now 'FMS fever' seems to be developing in Japan, with small companies wanting to buy FMS rather than an NC machine.

Now that Yamazaki and others have shown the advantages of using robots at machine tools a trend has been set, and use will accelerate. In many cases, the machine-with-robot will be installed because that is the trend, rather than because there is mathematical evidence of a big increase in productivity.

Soon a similar situation will develop in assembly, not least because assembly robots are becoming cheaper. The Scara breed is priced at around £10,000-£15,000 while Hitachi is introducing a simple assembly robot priced at about £2,500. By 1984 it is almost certain that many other companies will be selling small, simple robots for around £2,500. Production targets will have

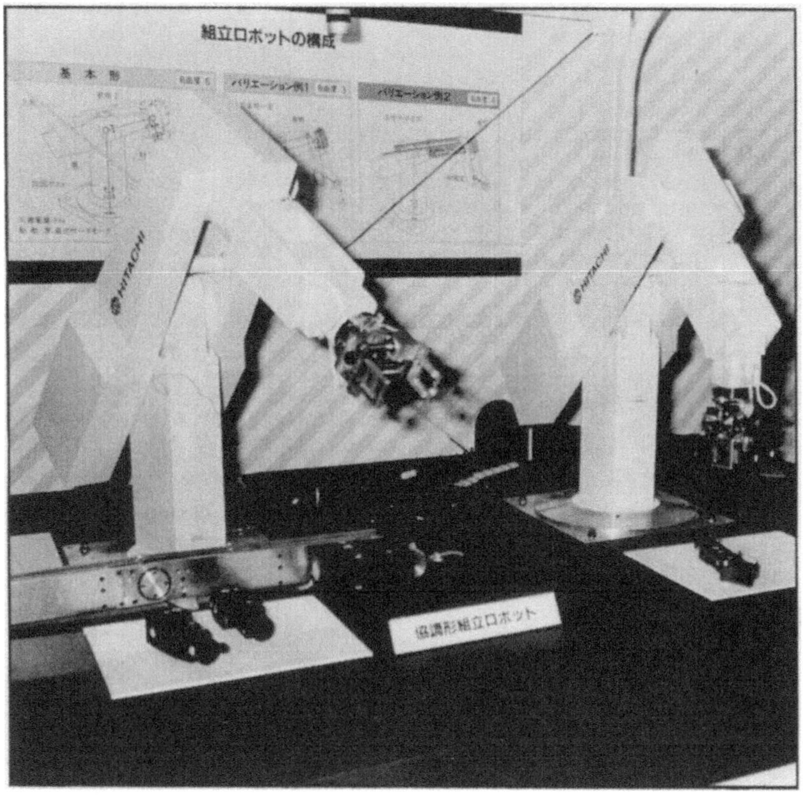

Fig. 5. Cheap assembly robots will help open the floodgates to robot assembly.

an aggregate of thousands rather than hundreds of machines a month.

Once a major manufacturer starts to install such robots in volume, then a whole host of companies will follow.

So why did the Fanuc assembly factory not inspire that breakthrough? In the first place, it is recognised that Fanuc can well afford to equip such a factory with robots for assembly to promote sales of robots. Secondly, the product is very specialised and Fanuc's competitors are generally large companies. The use will take off when it is clear that robots are economic in the assembly of products found in thousands of factories.

Official estimates put robot growth in Japan at about 20% a year for the next few years. Most manufacturers seem to agree with those figures although they expect the growth curve will have its stop and go periods.

In the next decade, it seems certain that robot use will spread through a wider range of jobs and into the small companies. Within a few years, we can expect almost all injection moulding machines to be sold equipped with robot arms. Many machine tools will be similarly equipped, but the larger ones will rely on separate robots. Equally, robots will be used far more in car body welding, pushing the robot welding level up from 50-75% to 75-90%.

Assembly is clearly on the threshold of a major expansion in robot usage, although this may take several years to materialise in a significant way. At the same time Japan's Ministry of International Trade and Industry (MITI) is sponsoring an ambitious project to develop an intelligent robot, while even more intelligent(!) robots are expected to emerge from the government-sponsored fifth-generation computer project.

By 1993 industrial robots will have become everyday machines that yield flexibility; but the signs are that the lion's share will be found in Japan. Some very complex robots will be used to supervise unmanned workshops, while robots will be used to inspect nuclear reactors and other dangerous areas as a matter of course.

'Can the Japanese overcome unemployment'

At the bottom end of the market the low cost of electronics will make it practical for any handling device to be programmable so that it can deal with many different products. Indeed, robots should be doing almost all jobs involving manual effort. The Japanese, who take such a long view on productivity and business gains but such a short view on the effects on society, are certain to be in the forefront as users of these new robots.

Can the Japanese find a way to overcome the problems of employment? The answer is not clear. In the past the Japanese answer always has been to increase GNP to mop up the extra workforce.

Even now the Japanese distribution system and retail outlets employ far more people than necessary by European standards, so there seems to be little scope for a wholesale switch from manufacture to the traditional services.

On the other hand, the emerging information industry, from computer software to data banks and multichannel television will generate major business which will certainly change the balance of society in such a way that worries about robots taking over from humans are likely to be groundless – so long as industry remains competitive, with good products and high productivity.

Whatever happens, it is difficult to see anything stopping the Japanese from pursuing their goals of higher productivity through the use of robots –wherever that may take them.

Why the robot will replace the Japanese navvie

By Professor Yukio Hasegawa of Waseda University.

WAVES of automation have been penetrating various industrial processes, ranging from the simple to the complex. Included in these is the industrial robot which has been helping to accelerate the diffusion of automation and bring breakthroughs into many difficult fields.

The most successful fields of robot application today are mainly in mass and medium scale manufacturing operations such as those found in building automobiles, electric appliances, electronics and so forth.

On the other hand even today many industrial operations remain in a poor state as a result of not receiving any of the benefits of innovative robot technology. For example, productivity in the construction industry is behind that found in advanced industries; indeed there exists a big gap between them.

For several years the author and a group of colleagues have been participating in a construction industry robot research project. Last year the group reformed under the name of WASCOR (Waseda Construction Robot) research project group. Recently the Japanese government also has begun provisional activities to study robot research in this field.

Fig. 1. Manual constructing operations.

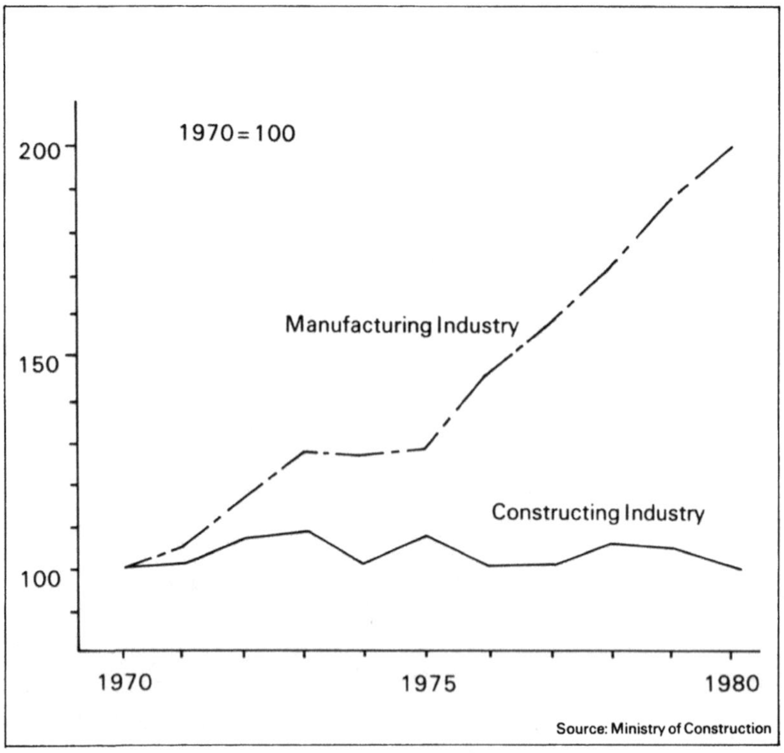

Fig. 2. Productivity increment trends in types of industry in Japan.

This article introduces a methodology for the research into such robots, examines problems in the construction industry, and looks at conceptional design of construction industry robots. Also shown are examples of robot development. Some of the problems still to be solved for the successful introduction of robots are discussed as well.

First let us look at some of the problems of introducing robots in the construction industry. Fig. 1 shows a typical scene involving manual construction operations. Today, the majority of concrete building construction work is done manually. The reasons why robotization has not yet been introduced are as follows:

- Buildings use a large number of different items.
- Product size is large and the weight of components is great.
- The location of the construction site is constantly changing.
- A variety of operations is needed to complete a building so multi-functions are needed for robots.
- There are many complicated group operations.
- Information feed-back from production people to product designers is poor; such people cannot improve the construction productivity because of many design restrictions.
- The stability of operation is inadequate.

Fig. 2 shows the recent tendency of productivity to increase throughout many types of industry. In the construction industry however there has been no increment of productivity rise in spite of the rapid increase in manufacturing industry.

Fig. 3 shows the trend of material, labour and total costs of wooden concrete form construction per square metre during the past 15 years. In this time labour costs have increased more than four times.

Because of the lower productivity

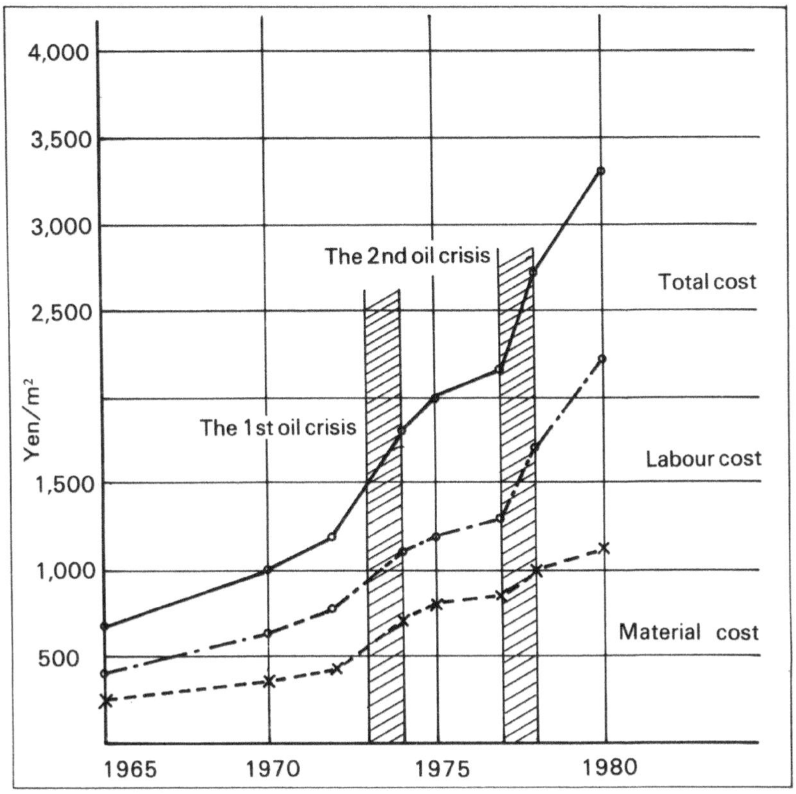

Fig. 3. Recent plywood form cost trends.

today costs are increasing yet on the other hand working conditions remain poor. For that reason people in this industry have begun to consider the introduction of robots to break through the thick wall of the seriously low productivity.

Turning now to the conceptual design of construction industry robots it is important to look at the methodology for introducing industrial robots into the industry. To conquer the difficulties of introducing robotization throughout general construction we have developed a process chart as shown in Fig. 4. The flow chart was used as a basis of the research work.

In the research emphasis was put on the following points:

- ☐ To maintain a consistent system approach from the beginning to the end.
- ☐ To make a survey of flow of technology and its forecasting.
- ☐ To decide the priority of operations to be robotized based on a questionnaire survey and an analysis of operations.
- ☐ To select an appropriate method of construction for robotization.
- ☐ To decide the robotized system targets to be accomplished as a result of the research.
- ☐ To conceive a building model as a foundation for the research.
- ☐ To determine the restrictions as a premise of conceptual design of the new robotized systems.
- ☐ To make a conceptual model design of the robotized building construction system.
- ☐ To determine the expected specification of robots and other associated components.
- ☐ To make an evaluation of the model alternatives.
- ☐ To make a feasibility study of models.
- ☐ To determine the problems for materializing model plans and future themes to be solved.

For this purpose a questionnaire was sent to operators, subcontractors, managements and general field supervisors in the construction industry. In addition using a memo-motion film analysis method, a study was made of operator posture and fatigue.

Fig. 5 shows the result of survey of operators' when handling reinforced steel bars during assembly work. Fig. 6 is an example of robotization of priority based on the above mentioned survey.

'Modifying conventional methods'

What should be the design targets for a construction industry robot? To achieve a break-through in solving this difficult problem it is necessary to decide some fundamental policies to modify the conventional process and methods of construction. For example, it is necessary:

- ☐ To develop a new construction process which will increase the ease with which robots can be introduced in place of conventional methods.
- ☐ To reject dangerous operations from the process or use robots as substitutes for operatives undertaking dangerous work.
- ☐ To standardise and modularise each component of the building and let the new system help introduce the advanced CAD-CAM system.
- ☐ To seek the possibility of introducing prefabrication production components and sub-assemblies.
- ☐ To remove disassembling operations from construction processes which are difficult to robotize.
- ☐ To consider adopting three types of robot (automatic, remote control and hybrid) because of the flow repeatability of construction work.

So how is the conceptual design of new building construction robot likely to look? Fig. 7 shows an artist's impression of a future robotized building construction yard. In the years ahead many types of robots will be used, including those for ground digging, soil handling, steel beam assembling, form assembling, reinforcing bars assembly, concrete pouring, painting, measuring and so

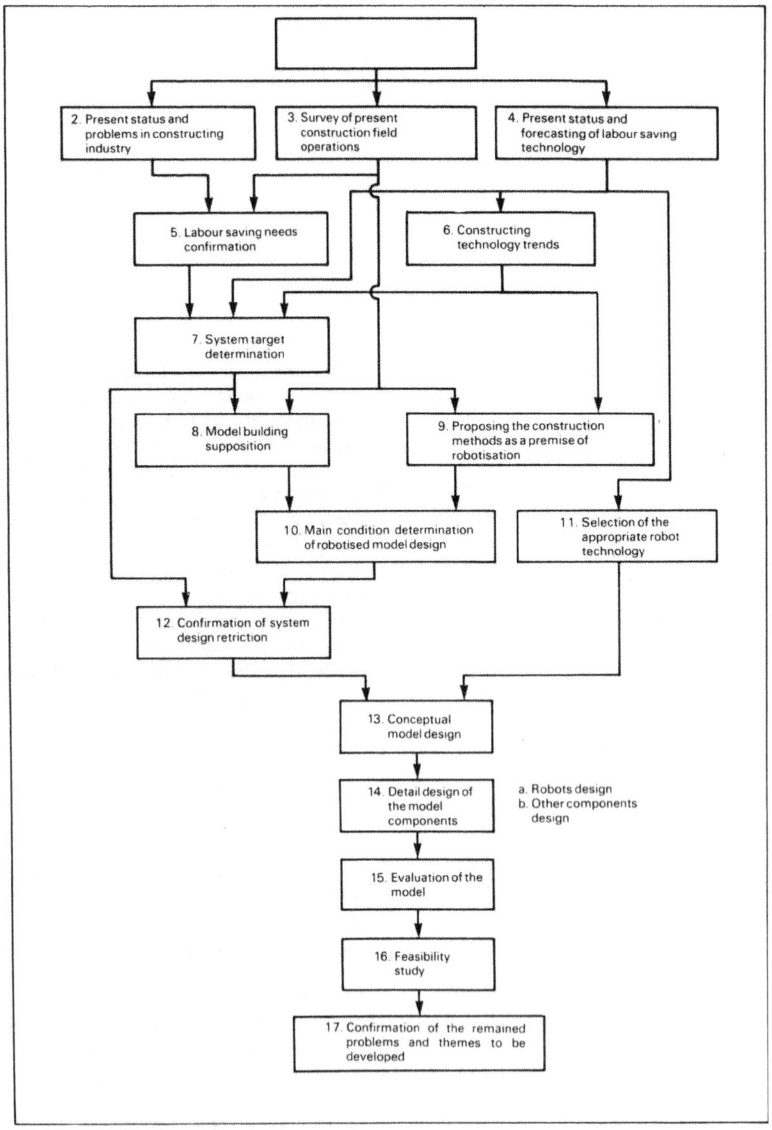

Fig. 4. Flow process chart of the research.

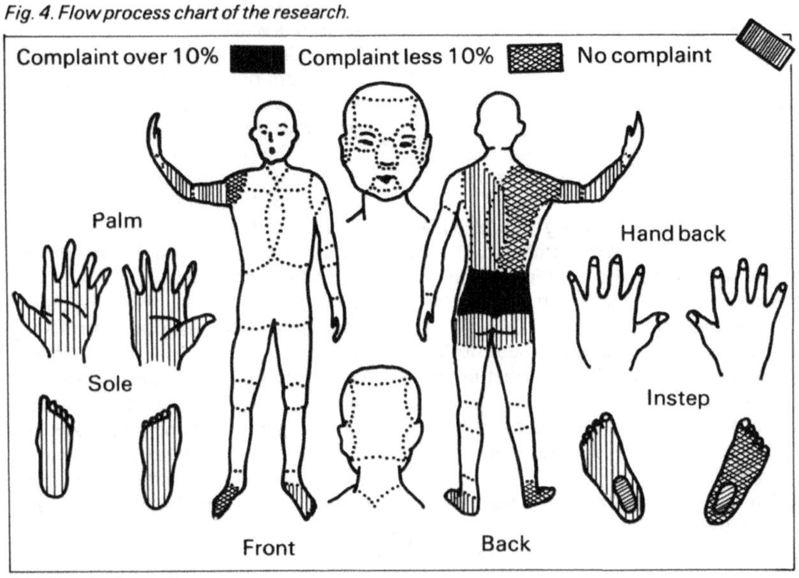

Fig. 5. Operators fatigue concentration.

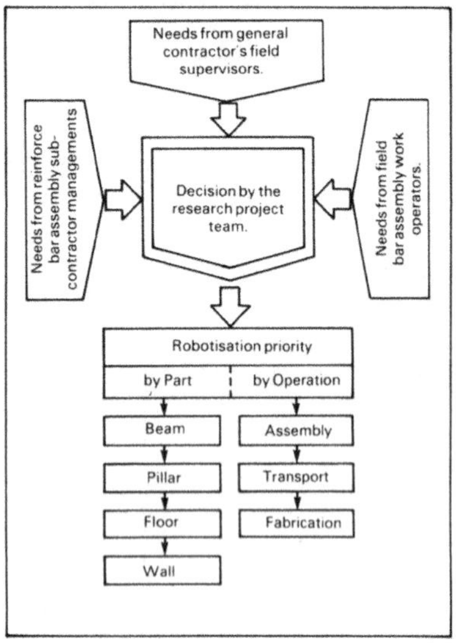

Fig. 6. Robotization priorities.

forth.

Building constructing operations are composed of dozens of complicated operations. It is therefore important to give a wide variety of functions to any robot used in the construction industry.

'The invasion of industrial robots'

For several years many leading Japanese construction firms and robot manufacturing companies have been continuing their efforts to develop construction robots for the sake of saving constructing cost, and improving safety and working conditions. This robotization is not yet totally synthesized in the way that factory automation is in manufacturing industry: designs shown in Figs. 8–11 suggest the future direction for robots in construction operations.

In conclusion it is worth pointing out that even today workers in the construction industry have problems of low wages, long working hours, and poor working environments because of lower productivity compared with other advanced industries. Whether we like to do so or not, people in the construction industry are expecting the invasion of

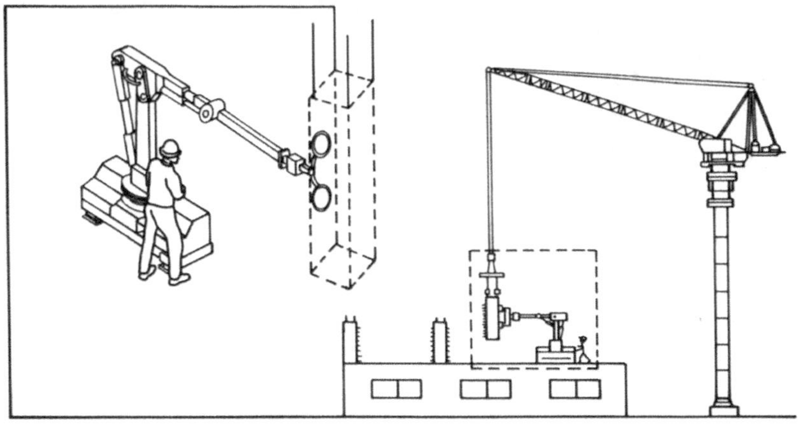

Fig. 7. Image of a robotized constructing yard.

Fig. 8. A concrete distributing robot.

Fig. 9. A stud bolt setting robot for atomic power plant building.

Fig. 10. An antifire insulating coat.

Fig. 11. Large concrete panel handling robot.

industrial robots and a solution to their problems by using advanced robot technology.

For fulfilling such social needs it is important to develop and apply more systematic methodology of industrial robot application. It is certainly worthwhile, taking painstaking steps in this area of robot research and development.

References

1. 'A Research Report on Robotization of Concrete Reinforce Bars Assembly Systems', Japan Industrial Robot Association, 1980, (In Japanese).
2. 'A Research Report on Robotization of Building Concrete Molds Handling and Assembly Systems', Japan Industrial Robot Association, 1981 (In Japanese).
3. Hasegawa Y., 'Workstudy for Robot Operation System Design', Proc. 8th ISIR (1978).
4. Hasegawa Y., 'Analysis of Complicated Operations for Robotization,' SME Tech. Paper (1979).
5. Hasegawa Y., 'Robotization of Reinforced Concrete Building Construction', Proc. 11th ISIR (1981).

AUTOMAN'83

Your stand by stand guide of exhibitors

A complete list of exhibitors, correct at the time of going to press, is shown below. Automan '83 takes place at the National Exhibition Centre, Birmingham, UK, 17-20 May 1983.

A

Aerotech Limited; 911
19A Livingstone Road
Newbury
Berkshire RG14 7PD
Contact: S. Savory
Tel: 0635 46313

Airstead Industrial Systems Ltd; 1605
New England House
New England Street
Brighton BN1 4GH
Contact: J. M. Simon
Tel: 0273 689793

Aoip Kremlin Robotique (AKR); 1725
ZI de Saint Guenault
6 rue Maryse Bastie
91031 Evry Cedex
France
Contact: J. W. Mathers
Tel: 0753 76013

Ariel Automatic Feeders; 1505
(Division of Ariel Industries PLC)
Ariel Works
Temple Road
Leicester LE5 4JG
Contact: D. W. Nunn
Tel: 0533 736541

Asea Limited; 1705
25 Alston Drive
Bradwell Abbey
Milton Keynes MK13 9HA
Contact: B. D. Sugg
Tel: 0908 319666

A. T. M. Engineering Ltd; 1501
Unit 9
Earls Way
Churchill Road
Thurmaston
Leicester
Contact: R. D. Hopper
Tel: 0533 693396

Automated Process Inc (A.P.I.); 1200
P O Box 23217
7869 North 73rd Street
Milwaukee
Wisconsin 53223–0217
U.S.A.
Contact: R. J. Lauer
Tel: (414) 354–4370

Automatix International France; 1604
3 Rue Troyon
75017 Paris
France
Contact: C. Baudier
Tel: 227 4972

Aylesbury Automation Ltd; 1225
P O Box 2
Mandeville Road
Aylesbury
Buckinghamshire HP21 8AB
Contact: C. A. Harrington
Tel: 0296 5911

B

Badalex Limited; 1615
Weybridge Trading Estate
Weybridge
Surrey KT15 2RH
Contact: N. D. Bell
Tel: 97 48311

BL Systems Ltd; 1733
BL Technology Ltd
P O Box 5
Redditch
Worcs
Contact: K. Humphries
Tel: 0527 64274

Bodine Products; 1535
Bridgeport Machines Division of Textron Limited
P O Box 22
Forest Road
Leicester LE5 0FJ
Contact: B. J. Hancock
Tel: 0533 531122

Bofors Electronics Ltd; 735
Murdock Road
Bedford MK41 7PQ
Contact: D. Viney
Tel: 0234 49241

E & E Bridges Limited; 1334
P O Box 36
Henley on Thames
Oxon
Contact: E. J. Bridges
Tel: 04912 6126

British Robot Association; 1215
35–39 High Street
Kempston
Beaford
Contact: T. E. Brock
Tel: 0234 855271

British Robotic Systems Limited; 1124
9/11 Grosvenor Gardens
London SW1W 0BG
Contact: B. M. Atkinson
Tel: 01–828 4140/4137

Norman Butter & Co Ltd; 1727
P O Box 100
Clooingwood Road
Coventry CV5 6HW
Contact: D. B. Pearson
Tel: 0203 711211

C

Capital (UE) Ltd; 1600
12 Tulip Tree Avenue
Kenilworth CV8 2BU
Contact: D. A. Hill
Tel: 0926 54293

Cincinnati Milacron Limited; 1003
Industrial Robot Division
Caxton Road
Bedford MK41 0HT
Contact: G. F. Jones
Tel: 0234 45221

Citroen Industrie U.K.; 919
Augusta House
Augusta Place
Leamington Spa CV32 5EL
Contact: P. B. Purdom
Tel: 0926 38872

Comau SpA; 715
Via Rivalta 30
10095 Grugliasco
Torino
Italy
Contact: Mrs. Calvi
Tel: 003911 33341

Contraves Industrial Products Ltd; 1004
Electronic Drive Systems Dept
Mansion Close
Moulton Park Industrial Estate
Northampton NN3 1RU
Contact: G. Lockyer
Tel: 0604 493201

Cramic Engineering Co. Ltd; 929
Beaumont Road
Banbury
Oxon OX16 7RD
Contact: G. Greenhalgh
Tel: 0295 51415

Cranfield Institute of Technology; 1805
Cranfield
Bedford MK43 0AL
Contact: Prof. K. Rathmill
Tel: 0234 750111

Anthony H. Croucher Ltd; 1803
Holybourne
N. Alton
Hants
Contact: A. H. Croucher
Tel: 0420 82868

Cybermatic Limited; 1006
Heming Road
Washford Industrial Estate
Redditch
Worcs
Contact: A. R. Waite
Tel: 0527 27882

Central Office of Information; 1533
Department of Industry
St Christopher House Annexe
Sumner Street
London SE1
Contact: Mr. Hammam
Tel: 01–928 2371

D

Dainichi Sykes Robotics Ltd; 803
Walton Summit Centre
Bamber Bridge
Preston PR5 8AE
Contact: I. D. Sutcliffe
Tel: 0772 322444

Desoutter Limited; 1606
319 Edgware Road
Colindale
London NW9 6ND
Contact: P. Tobitt
Tel: 01–205 7050

Development Board for Rural Wales; 1706
Ladywell House
Newtown
Powys
Wales
Contact: R. F. Ankers
Tel: 0686 26965

The Devilbiss Company Ltd; 1119
Ringwood Road
Bournemouth BH11 9LH
Contact: I. Bunker
Tel: 020 16 71111

Devon & Cornwall; 601
Development Bureau
County Hall
Truro
Cornwall TR1 3BB
Contact: N. Robertson
Tel: 0872 71324

Dowty Hydraulic Units Limited; 1626
Arle Court
Cheltenham
Glos GL51 0TP
Contact: P. J. Russell
Tel: 0242 21411

Dixon Automatic Tool Ltd; 1420
Unit 2
Shrub Hill Industrial Estate
Shrub Hill
Worcester WR4 9EL
Contact: D. Rushton
Tel: 0905 20955

Dynacast International Ltd; 1625
Tything Road
Arden Forest Industrial Estate
Alcester
Warwickshire
Contact: S. Bullen
Tel: 0789 763322

E

Engineering & Scientific Equipment; 1301
22–26 Mount Pleasant
Alperton
Wembley
Middlesex
Contact: J. McClaran
Tel: 01–903 4721

Epple-Buxbaum-Werke; 1632
Aktiengesellschaft
Buxbaumstrabe 2
A–4600 Wels
Austria
Contact: R. Puhringer
Tel: 07242 7391

Evershed Robotics Limited; 1635
Bridge Road
Chertsey
Surrey
Contact: D. O. Obersby
Tel: 09328 61181

E. W. Robotic Control; 1833
1 London Road
Horndean
Portsmouth P08 0BN
Contact: I. M. Williams
Tel: 0705 592500

F

Fairey Automation Limited; 1715
Techno Trading Estate
Bramble Road
Swindon
Wilts SN2 6HB
Contact: G. Brown
Tel: 0793 481161

Fata European Group Spa; 1403
S.S.no 24km 12
10044 Pianezza
(Turin)
Italy
Contact: A. Esposito
Tel: 011/967.33.33

Fawcett Engineering Ltd; 931
Dock Road South
Bromborough
Merseyside
Contact: G. Walters
Tel: 051 645 2050

Frazer-Nash Group Ltd; 1724
Dundee House
Bepten Road
Midhurst
Sussex GU29 9LZ
Contact: R. Jones
Tel: 073 081 6141

G

GDA Gesellschaft Fur Digitale Automation MBH; 1203
Am Bahnhof 5
8915 Fuchstal
W. Germany
Contact: J. C. Boulaire
Tel: 08243/2012

GKN Lincoln Electric Limited; 1617
Black Fan Road
Welwyn Garden City
Herts AL7 1QA
Contact: Mrs. S. McCullagh
Tel: 96 24581

General Automation; 1818
Sterling House
20 Station Road
Gerrards Cross
Bucks SL9 8EW
Contact: Mr. W. Wells
Tel: 49 89112

Grossenbacher Elektronik AG; 925
Ostrasse 25
CH–9006 St Gallen
Germany
Contact: M. Cataldo
Tel: 071 24 01 01

G. V. Planer Limited; 1201
Windmill Road
Sunbury-on-Thames
Middlesex
Contact: A. Thake
Tel: 09327 86262

Gwent County Council; 1619
Chief Executive Officer's Dept
County Hall
Cwmbran
Gwent NP44 2XH
Contact: D. J. Powell
Tel: 06333 67711

H

Hahn & Kolb (Great Britain) Ltd; 1511
6 Forum Drive
Leicester Road
Rugby
Warks CV21 1NY
Contact: M. T. Lockie
Tel: 0788 74261

Hall Automation Limited; 1211
Boughton Road
Rugby
Warks CV21 1BU
Contact: C. P. Knapman

Tel: 0788 2144

Harmo Robots Ltd; 1515
P O Box 20
Welwyn Garden City
Herts
Contact: M. E. Aldridge
Tel: 07073 32121

Harmonic Drive Ltd; 1008
5 Black Horse Way
Horsham
West Sussex RH12 1NU
Contact: D. S. Wheeler
Tel: 0403 56115

Hayes Pilgrim Automation; 927
51 Huntsman Lane
Stamford Bridge
York
Contact: A. Hayes
Tel: 0759 71001

Haynes & Fordham Ltd; 1024
Unit 1
Springfield Industrial Estate
Farsley
Leeds LS28 7UT
Contact: N. Haynes
Tel: 045 382 3611

Heidenhain (GB) Limited; 1525
200 London Road
Burgess Hill
West Sussex RH15 9RD
Contact: H. Behrens
Tel: 04446 47711

Helbar Automation Ltd; 1235
401/403 Rayleigh Road
Eastwood
Leigh-on-Sea
Essex SS9 5JG
Contact: K. E. Collins
Tel: 0702 522425/522575

Herga Electric Limited; 700
Northern Way
Bury St. Edmunds
Suffolk IP32 6NN
Contact: E. Dutnall
Tel: 0284 701422

Hirata Industrial Machineries Company Limited; 1607
5–4 Myotaiji
Kumamoto
Japan 860
Contact: T. Imamura
Tel: 0963 44 2266

HMK Technical Services Ltd; 610
River Dane Road
Eaton Bank Industrial Estate
Congleton
Cheshire CW12 1UG
Contact: M. Krajewski
Tel: 02602 79411

I

IMI Norgren Enots Ltd; 600
Norgren Works
Campden Road
Shipston-on-Stour
Warwickshire CV36 4PX
Contact: J. B. Smith
Tel: 0608 61676

IMO Precision Controls Ltd; 624
1000 North Circular Road
London NW2
Contact: J. Winter
Tel: 01–452 6444

Infranor Limited; 921
26 Lambourne Close
Crawley
West Sussex RH10 6JW
Contact: A. D. Jones
Tel: 0293 29467

Ingersoll Engineers; 1015
Bourton Hall
Rugby
Warks CV23 9SD
Contact: Mrs. C. Reeve
Tel: 0926 27088

Intergrated Photomatrix Ltd; 1101
The Grove Trading Estate
Dorchester DT1 1SY
Contact: I. Clark
Tel: 0305 63673

I.S.M. Equipments Industriels De Montage; 1704
28 CH Du Martinet
C.H. 1007 Lausanne
Switzerland
Contact: W. R. Davenport
Tel: 0524 761687

Italian Trade Centre; 1127/1227/1327
37 Sackville Street
London W1X 2DQ
Contact: G. Giordano
Tel: 01–734 2412

K

Kerry Ultrasonics Limited; 1321
Hunting Gate
Wilbury Way
Hitchin
Herts SG4 0TQ
Contact: G. E. Littleford
Tel: 0462 50761

Kollmorgan (UK) Ltd; 1319
Inland Drive Division
13 Newnham Street
Bedford MK40 3JR
Tel: 0234 211171

Kongsberg; 1224
Priors Way
Maidenhead
Berks SL6 2HP
Contact: T. Kitson
Tel: 0628 39292

George Kuikka Limited; 1219
Hill Farm Avenue
Leavesden
Watford
Herts WD2 7BL
Contact: D. M. Leaver
Tel: 09273 70611

Kuka Welding Systems & Robot Ltd; 819
7 Union Street
Bedford MK40 2SF
Contact: J. C. DeVries
Tel: 0234 212546

L

Lambda Photometrics Ltd; 1800
Lambda House
Batford Mill
Harpenden
Herts AL5 5BZ
Contact: R. Carless
Tel: 05827 64334

Lamberton Robotics Limited; 1519
26 Gartsherrie Road
Coatbridge ML5 2DL
Contact: D. S. Preston
Tel: 0236 26177

Lanco AG, Lanco Economic; 1719
Gurzelenstrasse 12
4512 Bellach
Switzerland
Contact: Ch. Ruprecht
Tel: 065 38 22 22

P. N. Langmead Ltd; 1100
Enterprise House
Ashford Road
Ashford
Middlesex
Contact: L. Hughes
Tel: 07842 46901

Lansing Industrial Robots; 1618
Kingsclere Road
Basingstoke
Hants RG21 2XJ
Contact: A. Travers
Tel: 0256 3131

Laser Lines Ltd; 1700
19 Best Bar
Banbury
Oxon OX6 9SA
Contact: M. Taylor
Tel: 0295 57581

L. K. Measurement & Control Systems Limited; 604

East Midlands Airport
Castle Donington
Derby DE7 2SA
Contact: Mrs. Hunt
Tel: 0332 811349

M

Machinoexport Ftec; 1627
5 Aksakov Str.
Sofia
Bulgaria

Mai United Kingdom Ltd; 1325
Black Arrow House
2 Chandos Road
London NW10 6NF
Tel: 01–221 1616

Manifold Industries Ltd; 1315
Seymour Road
Leyton
London E10 7LZ
Contact A. A. Quintin
Tel: 01–556 1834

Martonair Limited; 1614
St. Margarets Road
Twickenham TW1 1RJ
Contact: W. J. Huxtable
Tel: 01–892 4411

Marwin Production Machines; 612 & 915
Waddens Brook
Wednesfield
Wolverhampton
Contact: J. E. Nicklin
Tel: 0902 65363

Mecman Limited; 1527
Mecman House
Suttons Park Avenue
Reading
Berks RG6 1AZ
Contact: T. D'arcy
Tel: 0734 664456

Microbo S.A.; 1534
Avenue Beauregard 3
CH–2035 Corcelles
Switzerland
Contact: T. P. Daly
Tel: 038 31 57 31

Midland Electric Manufacturing Company Limited (Electronics DIV); 1601
Regency House
101 Hagley Road
Edgbaston
Birmingham B16 8LA
Contact: T. Beadman
Tel: 021 454 9679

Modular Automation Ltd; 1311
30–31 St. Georges Sq
Worcester WR1 1HX
Contact: D. J. Cockayne
Tel: 021 784 9050

Modular Robotic Systems Ltd; 1025
16 Waterloo House
East Street
Worcester WR1 1NF
Contact: E. V. Williams
Tel: 0905 20764

Moog Controls Limited; 1611
Ashchurch
Tewkesbury
Glos GL20 8NA
Contact: S. Foy
Tel: 0684 296600

Mouldmation Limited; 1115
Unit 1C
Chase Park Industrial Estate
Chase Terrace
Walsall WS7 8JQ
Contact: G. J. Webster
Tel: 05436 2075

N

NCR Limited; 1320
206 Marylebone Road
London NW1 6LY
Contact: Mrs. L. Coles
Tel: 01–388 8253

Nimak – M.A.G.; 1027
D–5248 Wissen
West Germany
Contact: Miss Stephan
Tel: 02742 4024

Nordson (UK) Limited; 1415
Ashurst Drive
Cheadle Heath
Stockport SK3 0RY
Contact: D. Wood
Tel: 061 492 2882

Norprint International Ltd; 1634
Horncastle Road
Boston
Lincolnshire
Contact: D. B. Riley
Tel: 0205 65161

O

Olivetti Controllo Numerico; 1703
Stradale Torino 603
10090 S Bernardo D'Ivrea (To)
Italy

Origa UK Limited; 1135
Ashchurch Industrial Estate
Tewkesbury
Glos GL20 8TD
Contact: R. Speller
Tel: 0684 298977

P

Paterson Production Machinery; 1028/1134
Bridge Road
Camberley
Surrey
Contact: I. Paterson
Tel: 0276 20915

Patscentre International; 1303
Back Lane
Melbourne
Nr. Royston
Herts SG9 6DP
Contact: S. Blake
Tel: 0763 61222

Henry Pels & Co Ltd; 1718
32–38 Osnaburgh Street
London NW1 3NE
Contact: C. C. Tauchert
Tel: 01–87 4113

Pera; 1735
Melton Mowbray
Leicestershire LE13 0PB
Contact: P. Tapsell
Tel: 0664 64133

Portescap (UK) Ltd; 1701
204 Elgar Road
Reading
Berks RG2 0DD
Contact: Mrs. M. Mortimer
Tel: 0734 861485

Portable Welders Ltd; 905
Churchill Road
Bicester
Oxfordshire OX6 7UP
Contact: I. C. Brent-Smith
Tel: 086923 688

Powertran Cybernetics; 933
Portway Industrial Estate
Andover
Hants SP10 3NN
Contact: H. Becker
Tel: 0264 64455

R

Rediffusion Simulation Ltd; 1532
Gatwick Road
Crawley
Sussex
Contact: B. Laven
Tel: 0293 35155

Regal Engineering (International) Limited; 1032
35 Spinney Hill
Addlestone
Surrey KT15 1AF
Contact: E. A. Heywood
Tel: 0932 48676

Ringway Power Systems Ltd; 1427
Churchill House
88–92 Talbot Road
Old Trafford
Manchester M16 0PD
Contact: Mrs. W. Jenkinson
Tel: 061 872 6829

Robot & Automation Consultants; 1010
Centec Engineering
17 Totman Crescent
Weir Industrial Estate
Rayleigh
Essex
Contact: B. Hawkins
Tel: 0268 741219

Robotec Limited; 1401
99 Derby Road
Nottingham NG1 5BB
Contact: M. Paradise
Tel: 0602 411600

Roevac Automation Limited; 1125
Grey Street
Denton
Mancheter M34 3RT
Contact: A. W. Watts
Tel: 061 336 3857

S

S.A.C. – Sceptre; 1033
Leyton Avenue
Mildenhall
Suffolk IP28 7BL
Contact: T. J. A. Brock
Tel: 0638 712776

Sciaky Electric Welding; 1514
212 Bedford Avenue
Slough (Trading Estate)
Berks SL1 4RH
Contact: F. E. Watkins
Tel: 0753 25551

Scicon Limited; 1425
49 Berners Street
London W1P 4AQ
Contact: R. M. Barker
Tel: 01–580 5599

Seagar Control Systems Ltd; 602
The Council Buildings
Spratton Road
Brixworth
Northampton
Contact: W. G. P. Jenner
Tel: 0604 880349

Siemens Limited; 1433
Eaton Bank
Congleton
Cheshire CW12 1PH
Contact: R. Caithness
Tel: 02602 78311

Sormel SA; 1314
Rue Becquerel
Z.I. De Chateaufarine
25000 Besancon
France
Contact: Mr. Marano
Tel: (81) 52.74.44.

Spemach Limited; 1603
Camphill Industrial Estate
Camphill Road
West Byfleet
Surrey
Contact: A. T. Pellins
Tel: 09323 40177

Stanton Automation; 1300
416 Wembley Commercial Centre
East Lane
Wembley
Middlesex HA9 7BG
Contact: M. D. Stanton
Tel: 01–908 1969

Syke Instrumentation Co. Ltd; 907
Syke House
117–119 Station Road
Liss
Hants GU33 7AJ
Contact: Mrs. S. Robey
Tel: 073 082 3821

Systems Control; 1234
Perry Avenue
Teeside Industrial Estate
Thornaby TS17 9LN
Contact: K. Carter
Tel: 0642 764823

T

Tandem Computers Limited; 1414
54 Hagley Road
Edgbaston
Birmingham B16 8PE
Contact: T. Hayward
Tel: 021 454 9772

Tayside Region Industrial Office; 1503
Tayside House
Crichton Street
Dundee DD1 3RB
Contact: H. J. Moody
Tel: 0382 23281

Telehoist Limited; 1435
Manor Road
Cheltenham
Glos GL51 9SH
Contact: R. G. Phillips
Tel: 0242 21355

Time & Precision (Sales) Ltd; 1500
Whitney Road
Daneshill East
Basingstoke
Hampshire RG24 0NS
Contact: R. S. Cox
Tel: 0256 28428

Toko (UK) Limited; 1002
Ward Royal Parade
Alma Road
Windsor
Berks SL4 3HR
Contact: R. W. Sanders
Tel: 07535 54057

Trallfa A/S; 1411
Robot Division
P O Box 113
4341 Bryne
Norway
Contact: J. Osthus
Tel: 04 482800

U

Unimatic Engineers Limited; 1801
122 Granville Road
London NW2
Contact: M. Stevens
Tel: 01–455 0012

Unimation (Europe) Limited; 1103
Unit C
Stafford Park 18
Telford
Shropshire
Contact: D. West
Tel: 0952 618931

Unimax Switch Ltd; 618
Martin Road
Cordwallis Industrial Estate
Maidenhead
Berks
Contact: M. J. Taylor
Tel: 0628 32646

V

Valley Automation Ltd; 835
Valley Road
Ley
Stourbridge
West Midlands
Contact: P. J. Smith
Tel: 038 482 2324

W

Welding Robotics & Automation; 1419
Units 2 & 3
Wulfrun Trading Estate
Stafford Road
Wolverhampton
Contact: E. A. Lane
Tel: 0902 711201

Welsh Development Agency; 1624
Treforest Industrial Estate
Pontypridd
Mid Glamorgan
Contact: Mr. L. Jenkins
Tel: 044 385 2666

Wickman Automated Assembly; 1019
Herald Way
Binley
Coventry CV3 2NY
Contact: J. Galvin
Tel: 0203 450810

X

X.D. Machine Tools Ltd; 903
Oaston Road
Nuneaton
Warwickshire CV11 6JX
Contact: G. J. Waters
Tel: 0203 347527

AUTOMAN'83

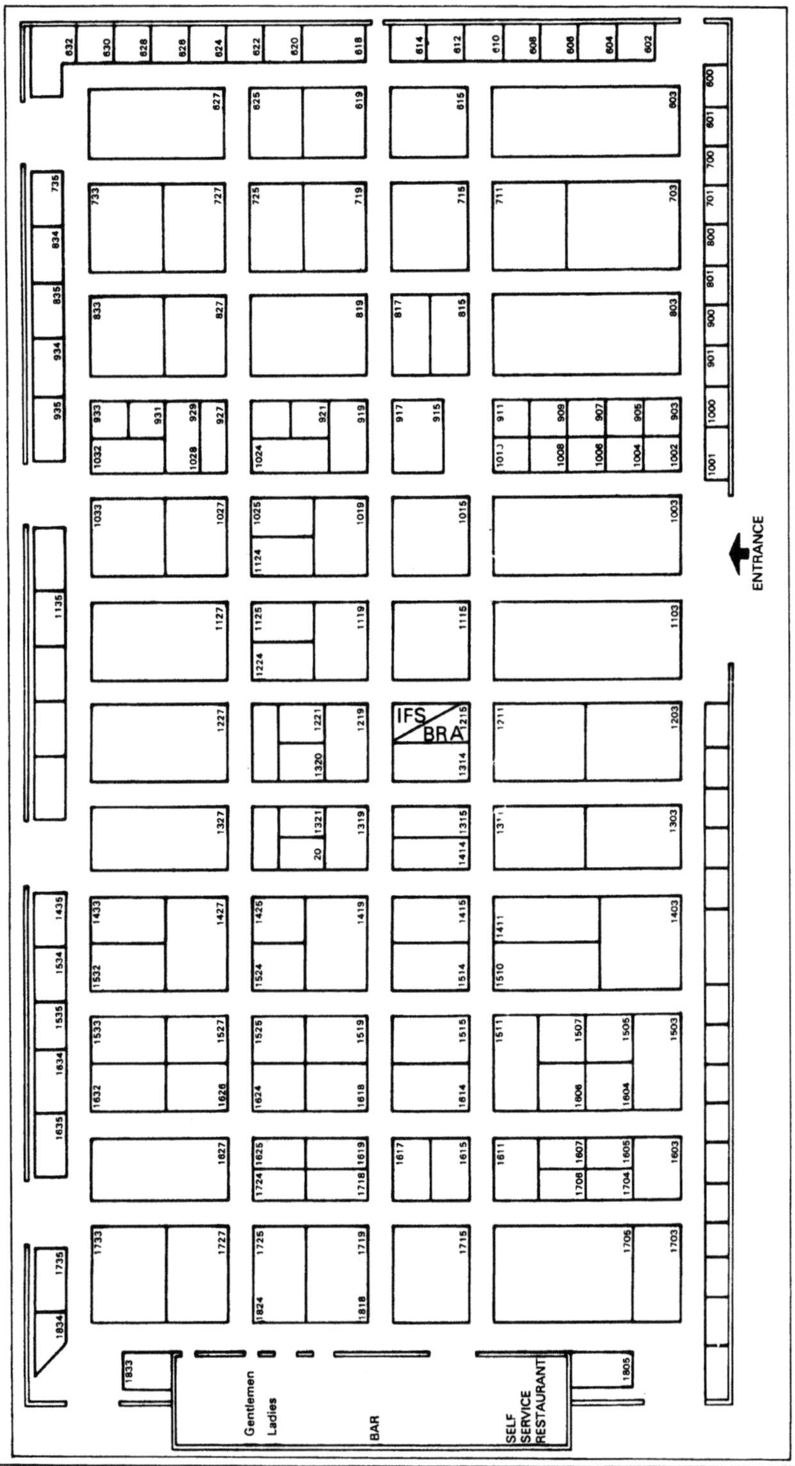

Stand number allocation at the NEC for Automan '83. IFS Publications is on Stand 1215. See us there.

ROBOTS 7

Robots 7 Exhibition in Chicago, USA, is the largest in the series yet staged by the organisers. Over 200 robots are 'in action' from a wide variety of firms whose names, addresses and stand numbers are given here Robots 7 is on 18-20 April 1983 at McCormick Place, Chicago, USA

A

A & A Manufacturing Co; 625
2300 S Calhoun Road
New Berlin
WI 53151
USA

A.S.R. Servotron, Inc; 241
2750 Oregon Court
M-3 Torrance
CA 90503
USA

Acrobe Positioning Systems, Inc; 167
Light duty assembly robots
3219 Dolittle Drive
Northbrook
IL 60062
USA

Action Machinery Co; 624
Forging and foundry manipulators/robots
PO Box 3068
Portland
OR 97208
USA

Admiral Equipment Co; 349
305 W. North Street
Akron
OH 44303
USA

Advanced Energy Technology; 725
839 Pearl Street
Boulder
CO 80306
USA

Advanced Robotics Corp; 125
Newark Industrial Park
8 Hebron
OH 43025
USA

Allen-Bradley Co; 220
747 Alpha Drive
Highland Heights
OH 44143
USA

Amcot Corp; 213
146-43 60th Avenue
Flushing
NY 11355
USA

Ameco Corp; 687
Slantbed lathe load/unload robot
PO Box 385
Menomonee,
WI 53051
USA

American Can Co; 291
Robotic components/parts, handling systems, turnkey system supplier
American Technologies Division
1900 Pollitt Drive
Fair Lawn,
NJ 07410
USA

American Robot Corp; 359
354 Hookstown Road
Clinton
PA 15026
USA

American SMT-Pullmax Inc; 669
900 E. Business Center Drive
Mount Prospect
IL 60007
USA

Andco Actuator Products, Inc; 719
4375 Walden Avenue
Lancaster
NY 14240
USA

Anorad Corp; 245
Robotic systems with positioning tables and computer controllers
110 Oser
Hauppauge
NY 11787
USA

Applied Robotics, Inc; 705
3 West Walnut Lane
Schenectady
NY 12309
USA

ASEA; 399
1176 E. Big Beaver Road
Troy
MI 48084
USA

Astek Engineering Inc; 721
Accommodaters, grippers and force sensors
5 Bridge Street
Watertown
MA 02172
USA

Atchley Controls Inc; 727
6933 Eton Avenue
Canoga Park
CA 91303
USA

Automated Process Inc; 732
Assembly machines
7869 N. 73rd Street
Milwaukee
WI 53223
USA

Automatic Systems; 148
330 W. 47th
Kansas City
MO 64112
USA

Automation Systems Inc; 728
Machine vision
1106 Federal Road
Brookfield
CT 06804
USA

Automation Systems Inc; 544
208 N. 12th Avenue
Eldridge
IA 52748
USA

Automatix Inc; 130
Vision systems
217 Middlesex Turnpike
Burlington
MA 01803
USA

Automaton Corp; 260
Industrial robots and part-handling devices
23996 Freeway Park Drive
Farmington Hills
MI 48024
USA

B

B.E.I. Electronics Inc; 101
681 Market Street
San Francisco
CA 94105
USA

Barrington Automation Ltd; 215
Pneumatic mini-robots and automation componentry
1002 South Road
Fox River Grove
IL 60021
USA

Bendix Corp; 345
Industrial robot systems
Robotics Division
21238 Bridge Street
Southfield
MI 48034
USA

Bink Manufacturing Co; 286
9201 W. Belmont Avenue
Franklin Park
IL 60131
USA

Alexander Binzel Corp; 158
Robot welding guns and components
3715 W. Ninth Street
Winthrop Harbor
IL 60096
USA

Bird-Johnson Co; 598
Rotary actuators, 3-axes robotic module
110 Norfolk Street
Walpole
MA 02081
USA

Black & Webster Inc; 645
281 Winter Street
Waltham

MA 02154
USA

Blanarovich Engineering; 592
P.O. Box 65
Don Mills
Ontario
Canada
M3C 2R6

C

Cincinnati Milacron; 400
Industrial robots
Industrial Robot Division
215 S. West Street
Lebanon
OH 45036
USA

Clifton Precision/Litton Systems Inc; 730
Marple & Broadway
Clifton Heights
PA 19018
USA

Cognex Corp; 495
Vision recognition system
1505 Commonwealth
Boston
MA 02135
USA

Comet Welding Systems; 398
880 Nicholas Boulevard
Elk Grove Village
IL 60007
USA

Compact Air Products; 413
Actuators for grippers, fixtures, tooling, pneumatic robot grippers
P.O. Box 176
Westminster
SC 29693
USA

Compumotor Corp; 718
1310 Ross Street
Petaluma
CA 94952
USA

Control Automation Inc; 467
Vision systems, 4-axes robots
PO Box 2304
Princeton
NJ 08540
USA

Copperweld Robotics Inc; 348
1401 E. 14 Mile Road
Troy
MI 48084
USA

Crawford Fitting Co; 275
29500 Solon Road
Solon
OH 44139
USA

Cybotech Corp; 671
Industrial robots
PO Box 88514
Indianapolis
IN 46208
USA

D

Delavan Electronics Inc; 128
14605 North 73rd Street
Scottsdale
AZ 85260
USA

Dependable-Fordath Inc; 355
Six-axes heavy-duty robot
400 S.E. Willamette Street
Sherwood
OR 97140
USA

Design Technology Corp; 221
53 Second Avenue
Burlington
MA 01803
USA

DeVilbiss Co; 424
P O Box 913
Toledo
OH 43692
USA

Diffracto Ltd; 186
Four-axes measuring robot with vision sensors
2775 Kew Drive
Windsor
Ontario
Canada N8T 159

Dojen Robotics Inc; 160
P O Box 1020
Woburn
MA 01801
USA

Dynamac Inc; 713
Industrial robots
410 Forest Street
Marlboro
MA 01752
USA

Dynamic Valves Inc; 295
923 Industrial Avenue
Palo Alto
CA 94303
USA

E

E & H Systems; 609
1907 Collegewood
Ypsilanti
MI 48197
USA

ERIM; 31
3300 Plymouth Road
P O Box 8618
Ann Arbor
MI 48107
USA

Eaton Corp; 525
EPP Division
Main & Orchard
Mantua
OH 44255
USA

Eaton-Kenway; 691
310 South Main Street
Salt Lake City
UT 84101
USA

EIC Inc; 615
48 W 38th Street
New York
NY 10018
USA

Emerson Electric Co; 171
Parts handlers, robots and components
Commercial Cam Division
1440 S. Wolf Road
Wheeling
IL 60090
USA

Equus Automated Systems; 267
3102 Kiowa Street
Fort Worth
TX 76114
USA

The Everclear Co; 722
3612 W. Lincoln Highway
Olympia Fields
IL 60461
USA

Everett/Charles Automation Modules Inc; 537
Assembly robots
6101 Cherry Avenue
Fontana
CA 92335
USA

EWAB/America; 298
Industrial robots, conveyors, turnkey applications
292 W. Palatine Road
Wheeling
IL 60090
USA

F

Fabrication Specialties; 566
4321 Taylorsville Road
Dayton

OH 45424
USA

Fairchild CCD Imaging; 235
3440 Hillview Avenue
Palo Alto
CA 94304
USA

Fared Robot Systems Inc; 599
Robotic automation of electronics assembly
3860 Revere Street
Suite D
Denver
CO 80233
USA

Ferguson Machine Co; 195
Division of UMC Industries Inc
11820 Lackland Road
St. Louis
MO 63141
USA

Festo Corp; 156
395 Moreland Road
Hauppauge
NY 11788
USA

Fibro Inc; 11
27 Franklin Avenue
Piermont
NY 10968
USA

Fleximation Systems Corp; 221
Toshiba Seiki robots for assembly applications
53 Second Avenue
Burlington
MA 01803
USA

Flow Systems Inc; 199
Nozzles for robots
21414 68th Avenue
S E Kent
WA 98032
USA

G

Gallaher Enterprises; 144
P O Box 10244
Winston-Salem
NC 27108
USA

GCA/Industrial Systems Group; 100
1 Energy Center
Naperville
IL 60566
USA

GDA/Automated Manufacturing Equipment Corp; 699
Luetzelsteiner Street 1
800 Muenchen 45
West Germany

General Electric Co; 300 & 320
One River Road
Schenectady
NY 12345
USA

General Electric Credit Corp; 471
260 Long Ridge Road
Stamford
CT 06902
USA

GMF Robotics Corp; 600
Robots, robotic systems
5600 New King Street
Troy
MI 48098
USA

Gould Inc; 596
Getty Motion Control Division
2701 N. Green Bay Road
Racine
WI 53404
USA

Graco Inc; 270
P O Box 1441
Minneapolis
MN 55440
USA

Greenwood Inc; 155
Ultra-Systems Division
75E Anchor Street
River Rouge
MI 48218
USA

H

Ham Industries Inc; 297
Vision systems
Inspection Products Division
835 Highland Road
Macedonia
OH 44056
USA

Harmonic Drive; 124
51 Armory Street
Wakefield
MA 01880
USA

Heathkit/Zenith; 639
Experimental robot-electronics education courses
Hilltop Road
St. Joseph
MI 49805
USA

Heidenhain Corp; 621
80 N. Scott Street
Elk Grove Village
IL 60007
USA

Hellstar Corp; 138
Robotic arm
1600 N. Chestnut
Wahoo
NE 68066
USA

Hirata Corp of America; 441
8900 Keystone Crossing
Indianapolis
IN 46240
USA

Hitachi America Ltd; 366
Five-axes process robot
Industrial Components
6 Pearl Court
Allendale
NJ 07401
USA

Hobart Brothers Co; 287
Robotic arc welding systems
600 W Main
Troy
OH 45373
USA

Hodges Robotics International Corp; 638
3710 N Grand River Avenue
Lansing
MI 48901
USA

Honeywell Motor Products; 215
4301 Kishwaukee Street
P O Box 106
Rockford
IL 61105
USA

E. F. Hutton Co; 233
200 Renaissance Center/Suite 2250
Detroit
MI 48242
USA

I

IBM Corp; 371
IRD Building 211/9B6,
PO Box 1328
Boca Raton
FL 33431
USA

IFS Publications Ltd; 113
Manufacturing technology books, magazines and newspapers
35 High Street
Kempston
Bedford
United Kingdom

Ikegai America Corp; 394
CNC turning machine with robot
2246 N Palmer Drive

Suite 108
Schaumburg
IL 60195
USA

Infranor Inc; 239 & 723
285 Murphy Road
Hartford
CT 06114
USA

Inland Motor; 655
Specialty Products Division
501 First Street
Radford
VA 24143
USA

Inoll Robotic Systems; 531
1411 Walnut Street
Suite 807
Philadelphia
PA 19102
USA

Intel Corp; 727
3065 Bowers
Santa Clara
CA 95051
USA

Intelledex Inc; 591
High-precision robot for electronics assembly
33840 Eastgate Circle
Corvallis
OR 97333
USA

International Robomation Intelligence; 370
6353 El Camino Real,
Carlsbad
CA 92008
USA

K

Keller Technology Corp; 271
Five-axes continuous path robots
Robotics Automation Systems
2320 Military Road
Tonawanda
NY 14150
USA

Kohol Systems, Inc; 571
PO Box 1185
Dayton
OH 45401
USA

Kollmorgen Corp; 225
Industrial Drives Division
201 Rock Road
Radford
VA 24141
USA

Kowa Trading Co. Ltd; 701
Robots and components
2-12-12-9FA Minamiaoyama
Minatoku
Tokyo 107
Japan

Kuka Welding Systems & Robot Corp; 448
24031 Research Drive
Farmington Hills
MI 48018
USA

Kurt Manufacturing Co; 187
5280 NE Main Street
Minneapolis
MN 55421
USA

L

Lord Corp; 708
Touch sensor
Industrial Products Division
1635 West 12th
Erie
PA 16512
USA

LTI Robotic Systems; 491
2701 Toledo Street
Suite 701
Torrance
CA 90503
USA

M

MTF Systems; 695
P O Box 24012
Minneapolis
MN 55424
USA

Machine Intelligence Corp; 449
330 Potrero Avenue
Sunnyvale
CA 94086
USA

Mack Corp; 159
Six-axes robot
3695 E Industrial Drive
Flagstaff
AZ 86001
USA

Manca Inc; 466
Robot systems and robot modules
PO Box 774
Menomonee Falls
WI 53051
USA

Marsden Electric Co; 217
PO Box 440
Howell
MI 48843
USA

Metalworking News; 663
7 E 12th Street
New York
NY 10003
USA

MFM Inc; 536
Modular robot
Flowerfield Industrial Park
Building 7
St James
NY 11780
USA

Michigan Dept of Commerce; 630
Office of Economic Development
525 W Ottawa Street
Lansing
MI 48909
USA

Microbot Inc; 266
453H Ravendale Drive
Mountain View
CA 94043
USA

Micron Instrument Corp; 24
210 Express Street
Plainview
NY 11803
USA

Missouri Research Laboratories; 533
630 Haines, NW
Albuquerque
NM 87102
USA

Mitsubishi Heavy Industries Ltd; 294
c/o Daiichi Jitsugyo American Inc
3158 Des Plaines Avenue
Des Plaines
IL 60018
USA

Modern Machine Works Inc; 183
Pick-and-place robotic loaders
Machine Tool Division
5355 S Kirkwood Avenue
Cudahy
WI 53110
USA

Moog Inc; 149
Industrial Division
Jamison Road
E Aurora
NY 14052
USA

N

National Robotics; 190
6 Mill Race
Spring House
PA 19477
USA

Nordson Corp; 200

Industrial coating robot; adhesive application robot systems
Robotics Division
555 Jackson Street
Amherst
OH 44001
USA

Northern Precision Laboratories; 715
11 Madison Road
Fairfield
NJ 07006
USA

O

Object Recognition Systems; 548
Vision system
1101 B State Road, Princeton
NJ 08540
USA

Octek Inc; 667
Vision system
7 Corporate Place
Burlington
MA 01803
USA

Omni Publications International Ltd; 292
909 Third Avenue
New York
NY 10022
USA

Origa Corp; 21
928 Oak Lawn Avenue
Elmhurst, IL 60126
USA

P

P H D Inc; 493
Modular robotic components such as grippers
Box 9070
Fort Wayne
IN 46899
USA

Parker Fluidpower; 176
17325 Euclid Avenue
Cleveland
OH 44112
USA

Pegasus Division; 40
2890 John R.
Troy
MI 48084
USA

Pendar/Kencotronics Inc; 540
Industrial robots
1821 University Avenue
St Paul
MN 55104
USA

Pentel of America; 257
2715 Columbia Street
Torrance
CA 90503
USA

Pertron Controls Corp; 588
21045 Osborne Street
Canoga Park
CA 91304
USA

PMI Motors; 253
Division of Kollmorgen Corporation
5 Aerial Way
Syosset
NY 11791
USA

Poly-Scientific; 177
Division of Litton Systems Inc
1213 N Main Street
Blacksburg
VA 24060
USA

Positech Corp; 638
Five-axes robot; manipulators
114 Rush Lake Road
Laurens
IA 50554
USA

Prab Robots; 500
6007 Sprinkle Road
Kalamazoo
MI 49003
USA

Precision Robots Inc; 261
6 Deangelo Drive
Bedford
MA 01730
USA

Productivity Systems Inc; 249
Robotic and automation systems, consulting
21999 Farmington Road
Farmington
MI 48024
USA

R

Reis Machines; 135
Electric robots
1426 Davis Road
Elgin
IL 60120
USA

Rexnord Inc; 649
Mechanical Power Division
P O Box 2022
Milwaukee
WI 53201
USA

Rob-Con Ltd; 586
Robots and robotic systems
Bra-Con Industries
12001 Globe Road
Livonia
MI 48154
USA

Robotic Systems Inc; 601
110 Technology Parkway
Norcross
GA 30092
USA

Robotic Vision Systems Inc; 445
Welding vision system
536 Broad Hollow Road
Melville
NY 11747
USA

Robotics Age; 722
174 Concord Avenue
Peterborough
NH 03458
USA

Robotics Inc; 594
RD No 3
Rt. 9
Ballston Spa
NY 12020
USA

Rolec Ltd; 535
Lease financing for industrial robots
3900 Chestnut Street
Suite 324
Philadelphia
PA 19104
USA

S

S I Handling Systems Inc; 545
PO Box 70
Easton
PA 18042
USA

Sandhu Rhino Robots; 635
Industrial robot, educational and research robot
308 S. State Street
Champaign IL
USA

Science Management Corp; 707
1120 Connecticut Avenue
NW Washington
DC 20036
USA

Scovill Automation Group; 390
Energy consultants in robots & industrial productivity
Schrader Bellows
200 W Exchange Street
Akron

OH 44309
USA

Seiko Instruments Inc; 552
2990 W Lomita Boulevard
Torrance
CA 90505
USA

Spectron Engineering Inc; 233
800 West 9th Avenue
Denver
CO 80204
USA

Sterling-Detroit Co; 499
261 E Goldengate
Detroit
MI 48203
USA

Strippit-Houdaille; 290
12975 Clarence Center Road
Akron
NY 14001
USA

Swanson-Erie Corp; 120-121
814 E 8th Street
Erie
PA 16512
USA

T

THK America Inc; 30
1715 Elmhurst Road
Elk Grove Village
IL 60007
USA

Tamagawa Seiki Co Ltd; 50
Rate and position sensors
3-19-9 Shinkamata
Ohta-Ku
Tokyo 144
Japan

Tarci Co; 109
Machine Design & Product Development
1180 Coleman Avenue
San Jose
CA 95110
USA

Tech Tran Corp; 710
134 N Washington Street
Naperville
IL 60540
USA

Technical Database Corp; 136
Industrial robot and automation database
Robotics Database Division
22916 Lyons Avenue
Suite 1103
Newhall
CA 91321
USA

Thermwood Corp; 324
Industrial robots
1910 Firman Drive
Suite 110
Richardson
TX 75081
USA

Tokico America Inc; 570
Painting robots and accessories
3555 W Lomita Boulevard
Suite E
Torrance
CA 90505
USA

Towa Corporation of America; 170
PO Box 513
Yardley
PA 19067
USA

U

Unimation Inc; 425
Shelter Rock Lane
Danbury
CT 06801
USA

United States Robots Inc; 244
Industrial robots
1000 Conshohocken Road
Conshohocken
PA 19428
USA

United Technologies Automotive Group; 391
5200 Auto Club Drive
Dearborn
MI 48126
USA

V

VSI Automation Assembly Inc; 166
Assembly robot
165 Park Street
Troy
MI 48084
USA

W

Warner Electric Brake & Clutch Co; 105
Gripper sensors
449 Gardner Street
S Beloit
IL 61080
USA

Westinghouse Electric Corp; 549
Westinghouse Building
Room 785
Pittsburgh
PA 15222
USA

Barry Wright Corp; 633
Components and accessories for robots
Barry Controls Division
700 Pleasant Street
Watertown
MA 02172
USA

Y

The Yankee Group; 566
89 Broad Street
Boston
MA 02110
USA

Yaskawa Electric America Inc; 248
Industrial robots
Motoman (Robots) Division
305 Era Drive
Northbrook
IL 60062
USA

The world's robot makers and suppliers

Robot manufacturers have struck up an interesting number of cross licencing deals with their agents, as these tables show.

A

Accumatic Machinery Corp 3537 Hill Avenue Toledo Ohio 43607, USA *(Tel: 419 535 7997)*		
ACMA-Cribier 3–5 rue Denis Papin 95250 France *(Tel: 3/4135490)*	US: *Cybotech Corp* *P.O. Box 88514* *Indianapolis, In. 46208* *(Tel: 317 298 5136)*	SPAIN: *Echazarra SA*
Acrobe Positioning Systems Inc 3219 Dolittle Drive Northbrook ILL 60062 USA		
Action Machinery Co PO Box 3068 Portland OR 97208 USA		
Admiral Equipment Co Ltd 305 West North Street Akron Ohio 44303, USA *(Tel: 216 253 1353)*	*See Yaskawa*	
Advanced Robotics Corp Route 79 Newark Industrial Park Hebron, Ohio 43025, USA *(Tel: 614 929 1065)*	*See Nachi Fujikoshi*	
Afma Robots BP 315, St Avertin 37173 Chambray les Tours Cedex France *(Tel: 47/276066)*		
Aida Engineering Ltd Automatic Machine Dept No 2–10 Oyama-cho, Sagamihara-Shi Kanagawa-Ken 229, Japan *(Tel: 0427 (72) 5231)*	BENELUX: *Chori GmbH* *Graf Adolfstrasse 59* *4 Dusseldorf* *West Germany*	SCANDINAVIA: *Chori Co Ltd* *Jernbanegade 17* *Copenhagen* *4 DK-1608 Denmark*
Airstead Industrial Systems Ltd New England House New England Street Brighton BN1 4GH, UK *Tel: 0273 689793*	*See* *Hirata* *Seiko* *Instruments*	
AISA via Roma 20 26020 Cumigano Italy		
Ajax Machine Tool Ltd CNC Machine Tool & Robot Division Knighton Heath Estate 847/855 Ringwood Road Bournemouth BH1		

AKR (Aoip Kremlin Robotique)
6 rue Maryse Bastie
ZI de Sait-Guenault
91031 Evry Cedex, France
(Tel: 6/077 9615)

UK:
Kremlin Spray Painting Equipment Ltd
839 Yeovil Road, Slough
Berks SL1 4JA, UK
(Tel: 0753 38143)

ITALY:
Kremlin
Via dei Lavoratori 116
120092 Cinisello Balsano
Balsano

GERMANY:
Kremlin Lackiersysteme
Phillips-Reis-Strasse 10
D6057 Deitzenbach
RFG

Ameco Corporation
PO Box 385
Menomonee
WI 53051
USA

American Robot Corp
201 Miller Street
Winston-Salem
North Carolina 27103, USA
(Tel: 919 748 8761)

UK:
Rediffusion Simulation
Gatwick Road
Crawley, West Sussex

Anorad
110 Oser Avenue
Hauppauge
New York 11788, USA
(Tel: 516 231 1990)

Ansaldo SpA
viale Sarca 336
20126 Milan
Italy

Armax Robotics Inc
38700 Grand River Avenue
Farmington Hills
Michigan 48018, USA
(Tel: 313 478 9330)

ASEA AB
Industrial Robot Division
S 72183 Vasteras
Sweden
(Tel: 021 100000)

UK:
ASEA Ltd
25 Alston Drive, Bradwell Abbey
Milton Keynes, UK
(Tel: 0908 319666)

FRANCE:
22 rue du 8 mai 1945
95340 PERSAN
(3) 470-92-00

SPAIN:
ASEA-CES
Apartado 81
Sabadell, Barcelona
(Tel: (93) 710 3800)

USA:
ASEA Inc
1176 E. Big Beaver
Troy, Michigan 48084, USA
(Tel: 313 528 3630)

ASEA AB (Previously Electrolux)
Industrial Robot Division Stockholm
Fagelviksvagen 3
S-145 53 Norsborg, Sweden
(Tel: 046 753 89100)

UK:
George Kuikka Ltd
Hill Farm Avenue
Leavesden, Watford, Herts
(09273) 70611

Atlas Copco
S-105 23
Stockholm

UK:
Atlas Copco GB Ltd
Swallowdale Lane
Hemel Hempstead, Herts.
(Tel: 0442 61201)

FRANCE:
Atlas Copco France
326 rue du General Heclere
95130 Franconville
(Tel: (3) 413 5454)

Automation Corporation
23996 Freeway Park Drive
Farmington Hills
MI 48024
USA

Automatix Inc
1000 Tech Park Drive
Billerica
Mass 01821, USA
(Tel: 617 667 7900)

FRANCE:
Automatix International France
3 rue Troyon
F 75008 Paris

JAPAN:
Tokyo Electron

ATM Engineering Ltd
Unit 9, Earls Way
Church Hill Ind Est
Thurmaston, Leicester LE4 8DH
(Tel: 0533 693396/7)

B

Barrington Automation Ltd
1002 South Road
Fox River Grove
IL 60021
USA

Basfer SpA — *See Nordson Corp*
via Iseo 60
20052 Monza
Italy

Bendix Robotics Division — *See Yaskawa*
21238 Bridge Street
Southfield
Michigan 48034, USA
(Tel: 313 352 7700)

Bertin & Cie
B.P. 3
78370 Plaisir
France
(Tel: 3/056 25 00)

Binks Bullows Ltd — *See GEC Robot Systems*
Pelsall Road
Brownhills, Walsall
West Midlands WS8 7HW, UK
(Tel: 05433 2571)

Binks Corp
9201 W. Belmont Ave
Franklin Park, Ill. 60131
(Tel: 312 671 3000)

Robert Bosch GmbH — *See Marwin Production Machines*
Geschaftsbereich Industrieausrustung
7000 Stuttgart 30, Kruppstrasse 1
Postfach 300220
(Tel: 0711 811 5225)

British Federal Welder and Machine Co Ltd
Castle Mill Works
Birmingham New Road
Dudley, West Midlands DY1 4DA, UK
(Tel: 0384 54701)

Norman Butters Co Ltd — *See Osaka Transformer Co Ltd*
P.O. Box 100
Collingwood Road
Coventry CV5 6HW
(Tel: 0203 711211)

C

Camel Robot srl
p.zza Addolorata 5
20030 Palazzolo Milanese
Milan
Italy

UK:
Fairey Automation
Techno Trading Estate, Bramble Road
Swindon, Wilts, SN2 6WB, UK
(0793) 481161

Can Engineering Sales — *See Prab Robots*
Pendar Robotics
P.O. Box 428
6800 Montrose Road
Niagara Falls,
Ontario, Canada

C.G.M.S. 98 rue D'Ambert B.P. 1825 45008 Orleans Cedex, France *(Tel: 38/86 25 14)*	*See Sankyo Seiki* *Taiyo* *Toshiba Seiki*		
Carl Cloos Schweisstechnik GmbH D-6342 Haiger West Germany *(Tel: 02773850)*	UK: *WRA*		
Ceeris International Inc 1055 Thames Jefferson St NW Ste 414 Washington DC 20007 *(Tel: 202 342-5400)*	*See Sormel SA*		
Chori	*See Aida Engineering Ltd*		
Cincinnati Milacron Industrial Robot Division 215 S. West Street, Lebanon Ohio, 45036, USA *(Tel: 513 932 4400)*	UK: *Cincinnati Milacron Ltd* *Robot Division, Dells Lane* *Biggleswade, Beds SG18 8LB, UK* *(Tel: 0767 313504)*	FRANCE: *Cincinnati Miliacron* *Division Marketing France, Silic 246* *94568 Rungis Cedex, Bureaux 2* *Place Gustave Eiffel (Tel: 687-22-44)*	
Cirrus Equipment Ltd Heming Road Redditch, Worcs B98 0DN *(Tel: 0527 27882)*			
Citroen 133 quai Andre Citroen 75747 Paris, Cedex 15 France *(Tel: 578 61 61)*	UK: *Citroen Industrie* *Augusta House* *Leamington Spa*		
Comau SpA via Rivalta 30 10095 Grugliasco (Turin) Italy	SPAIN: *Comau Iberica SA*	USA:	
Colne Robotics Ltd Beaufort Road, Richmond Road East Twickenham Middlesex, UK *(Tel: 01 892 8197)*	EUROPE: *Rapid Services System* *25a Hermitage Road* *London N4* *(Tel: 01-803 1961*	WEST GERMANY: *Dip Img Theobald* *Sthleiermacherstrasse* *D-6100 Darmstadt* *(Tel: 010 496151 26931)*	FAR EAST: *ELE International Ltd* *Eastman Way* *Hemel Hempstead, Herts, UK* *(Tel: 0442 50221)*
Comet Welding Systems 900 Nicholas Road Elk Grove Village Illinois 60007, USA *(Tel: 312 956 0126)*			
Commercy 55200 Commercy France *(Tel: (29) 010104)*	*See Shin Meiwa*		
Compania Anonima de Electrodos Infanta Carlota 56 Barcelona Spain *(Tel: (93) 666 50152)*	*See Yaskawa*		

Concentric Production Research Ltd
Upper Holland Road
Sutton Coldfield
West Midlands B72 1RD
(Tel: 021 355 1266)

Continental Parker
51 rue Pierre
92110 Clichy
France
(Tel: 793 3330)

Control Automation Inc
PO Box 2304
Princeton
NJ 08540
USA

Copperweld Robotics Inc
1401 E. Fourteen Mile Road
Troy
Michigan 48084, USA
(Tel: 313 585 5972)

Cybotech Corp
P.O. Box 88514
Indianapolis
Indiana 46208, USA
(Tel: 317 298 5136)

See Acma-Cribier

Cyclomatic Inc
7520 Convey Court
San Diego, Ca 92111
USA
(Tel: 714 292 7440)

See Thermwood Machinery

D

Daido Steel Co Ltd
7–13 Nishi Shinbashi 1-chome
Minato-ku, Tokyo, Japan
(Tel: 03/501 5261)

US:
Daido Steel
Suite 342 East, Pan American Bldg
200 Park Avenue
New York NY 10017

Daikin Kogyo Co Ltd
Laboratory, 700–1 Hitotsuya
Settsu City, Osaka
Japan
(Tel: 06/349 7361)

Daini Seikosha

See Seiko Instruments Inc

Dainichi Kiko Co Ltd
Kosiacho Nakakoma-gun
Yamanashi Prefecture
Japan
(Tel: 05528/2 5581)

UK:
Dainichi-Sykes
see below

US:
GCA/PAR Systems
3460 Lexington Avenue N
St Paul, Minnesota 55112
(Tel: 612 484-7261)

US:
Wamac Machinery Inc
9780 NW 79th Avenue
Hailech Gardens
Florida 33016

Dainichi-Sykes Robotics
Walton Summit Centre
Bamber Bridge, Preston
Lancs PR5 8AE, UK
(Tel: 0772 322444)

See Dainichi Kiko

Dale Electric of Great Britain Electricity Buildings Filey, North Yorks YO14 9PJ *(Tel: 0723 514141)*	*See Shin Meiwa*		
DEA SpA Corso Torino 70 10024 Moncalieri (Turin) Italy	UK: *Fairey Automation* *Techno Trading Estate, Bramble Road* *Swindon, Wilts, SN2 6WB* *(Tel: 0793 481161)*	SPAIN: *Oficina Tecnica Commercial*	US: *General Electric Automation Systems*
Dependable-Fordath Inc 400 SE Willimette St Sherwood OR 97140 USA			
DeVilbiss Company 837 Airport Boulevard Ann Arbor, Michigan, 48104, USA *(Tel: 313 668 6765)*	UK: *DeVilbliss Co Ltd* *Ringwood Road* *Bournemouth, Hants, UK* *(Tel: 0202 571111)*	FRANCE: *Toussand DeVilbiss* *163-171 ave des Aureats* *26014 Valance Cedex*	*See also Trallfa*
Diffracto Ltd 2775 Kew Drive Windsor, Ontario Canada N8T 159			
Dimenco PSP 16 rue Gay Lussac 25000 Besancon France *(Tel: 81/53 81 32)*	*See Novda*		
Distribel 33 rue Godwin, Ensival B-4850 Verviers Belgium *(Tel: 087 33 11 56)*	*See Ebosa*		
DOGA avenue Gutenburg B.P. 53, 78311 Maurepas Cedex France *(Tel: 3/062 41 41)*	*See Ebosa*		
Dubus 40 rue marceau 93100 Montreuil France *(Tel: 859 51 84)*			
Duffour et Igon rue de l'Oasis 31300 Toulouse France *(Tel: 61/42 35 36)*			
Duplomatic SpA via Alba 18 21052 Busto Arsizio (Varese) Italy	FRANCE: *Duplomatic Automation* *434 ave Pasteur* *94250 Gentilly*	US: *Duplomatic Systems Inc* *Los Verdes Ind Park.* *2045 South Valentia* *Denver, Colorado 80231*	
Dynamcac Inc 410 Forest Street Marlboro MA 01752 USA			

E

Ebosa
Kapellstrasse 26
CH 2540 Grenchen
Switzerland

ELAC
(See Ottensener Eisenwerk GmbH)

Electrolux AB
(See ASEA AB, Stockholm)

UK:
George Kuikka Ltd
Hill Farm Avenue
Leavesden, Watford, Herts
(Tel: 09273 70611)

Elicon
273 Viking Avenue
Brea
Calif 92621, USA
(Tel: 714 990 6647)

Emerson Electric Co
Commercial Cam Drive
1440 S Wolf Road
Wheeling, IL 60090
USA

H. Ernault-Somua
32 avenue de l'Europe
78140 Velizy-Villacoublay
France
(Tel: 946 96 40)

ESAB AB
Box 8004
S–40277 Gothenburg
Sweden

UK:
ESAB Ltd
Beeching Way, Gillingham
Kent ME8 6PU, UK
(Tel: 0634 34455)

FRANCE:
ESAB SA
176 ave Louis Roche
BP.81–92233 Gennevilliers Cedex
(Tel: (1) 799-98-46)

SPAIN:
ESAB Iberica SA
Zona Industrial
de Alcobendas
Madrid

USA:
ESAB North America Inc
1941 North Parkway
P.O. Box 2286, Fort Collins,
Colorado
(Tel: 303 484 1244)

Everett/Charles
Automation Modules Inc
9645 Arrow Route, Suite A
Rancho Cucamonga, Calif 91730
(Tel: 714 980 1525)

UK:
Everett/Charles (UK) Ltd
56 Wilbury Way
Hitchin, Herts SG4 0TP
(Tel: 0462 57577)

Evershed Robotics Ltd
Bridge Road
Chertsey
Surrey KT16 8LJ
(Tel: 09328 61181)

See Toshiba

EWAB America
292 W Palatine Road
Wheeling
IL 60090
USA

F

Fairey Automation Ltd Techno Trading Estate Bramble Road, Swindon Wilts SN2 6HB, UK *(Tel: 0793 481161)*	*See Camel* *DEA SpA* *Jobs SpA* *Pragma, Elfin,* *Gaiotta, Oglaend*	
Fared Robotic Systems Inc 3860 Revere Street, suite D P.O. Box 39268 Denver, Colorado 80239, USA *(Tel: 303 371 5868)*		
FATA-Bisiach & Carru Strada Statele 24, 12km 10044 Pianezza (Turin) Italy	UK: *FATA UK* *Dawson Road* *Milton Keynes*	
Gebr. Felss 7535 Konigsbach Stein 2 West Germany	FRANCE: *Felss France* *8 rue Commerce* *68400 Reidisheim Cedex 203, France.* *(Tel: 89/64 22 11)*	
Fibro GmbH Postfach 1120 D 6954 Hassmersheim West Germany	FRANCE: *Fibro France* *94–96 rue de la Goutte d'Or* *03300 Aubervilliers, France* *(Tel: 352 14 30)*	
Fleximation Systems Corporation 53 Second Avenue Burlington MA 01 803 USA	*See Toshiba*	
FN Robotics SA Shell Building 60 rue Ravenstein B-1000 Brussels, Belgium *(Tel: 02 511 2500)*	WEST GERMANY: *FN Robotics* *Voltastrasse* *D 6072 Dreilich* *Sprendingren*	*See Prab Robots*
France Euromatic 8 rue du Commerce 68400 Reidisheim France *(Tel: 89/64 15 33)*		
Frazer Nash Vine House 143 London Road Kingston-upon-Thames Surrey KT2 6NW		
Fuji Electric Co Ltd 12–1 Yuraku-cho 1 chome Chiyoda-ku, Tokyo Japan *(Tel: 03/211 7111)*		
Fuji Electric Co Ltd Hibiya Park Building, 8–1 Yurakucho 1-chome, Choyoda-ku, Tokyo Japan *(Tel: 03-201 1971)*	*See* *Haynes & Fordham*	

Fanuc Ltd Engineering Administration Dept 5–1 Asahiugaoka 3-chome Hino City, Tokyo, Japan *(Tel: 0425/84 1111)*	UK: *600 Fanuc Robots Ltd* *Colchester* US: *See GMF Robotics Corp*	FRANCE: *Halbroun Freres* *4 Rue Jean Moulin* *94130 Nogent-sur-Marne* *(Tel: (1) 873-69-45)*	WEST GERMANY: *Siemens AG* *Postfach 103* *D 8000 Munich 1*	US: *General Numeric Corp* *390 Kent Avenue* *Elk Grove Village, Illinois 60007* *(Tel: (312) 640-1595)*

G

Gaiotto Impianti SpA Statele Milano-Crema km 27 26100 Vaiano Cremasco (Cremona) Italy	UK: *Fairey Automation*	
GCA/PAR Systems 3460 Lexington Avenue, North St Paul Minnesota 55112, USA *(Tel: 612 484 7261)*	*See Dainichi Kiko*	
GEC Robot Systems Ltd Boughton Road Rugby CV21 1BD UK *(Tel: 0788 2144)*	UK: *Binks Bullows Ltd* *Pelsall Road, Brownhills, Walsall* *West Midlands WS8 7HW, UK* *(Tel: 05433 2571)*	*See Hitachi*
General Electric Automation Systems 1285 Boston Avenue, Bridgeport Connecticut 06602, USA *(Tel: 203 382 2876)*	*See DEA* *Hitachi* *Volkswagen*	
General Motors	*See GMF Robotics*	
G.D.A. 5 Am Bahnhof D-8915 Fuchstal West Germany *(Tel: 08243 2012)*		
General Numeric Corp 390 Kent Avenue Elk Grove Village Illinois 60007, USA *(Tel: 312 640 1595)*	*See Fanuc*	
George Kuikka Ltd Hill Farm Avenue Leavesden, Watford Herts, UK *(Tel: 09273 70611)*	*See ASEA* *Electrolux*	
GKN Lincoln Electric	*See Lincoln Electric*	
GMF Robotics Corp 5600 New King Street Troy MI 48098 USA	*See Fanuc*	

Graco Robotics Inc 12899 Westmore Ave Livonia Mi 48150 *(Tel: 313 261 3270)*			
Grundy and Partners Bond Mill Stonehouse Gloucestershire, UK *(Tel: 0453 823611)*	*See Haynes & Fordham*		

H

Haden Drysys Swan Centre 1056/8 Coventry Road Yardley, Birmingham, B25 8AD *)tel: 021-706 1623)*			
Hahn & Kolb Ltd 6 Forum Drive Leicester Road Rugby CV21 1NY, UK *(Tel: 0788 74261)*	*See* *ZF*		
Hall Automation Ltd *(See GEC Robot Systems Ltd)*			
Harmo Japan 7621-10 Fujizuka Nisha-minoco, Ina-City Nagano Prec *(Tel: 399-45)*	UK: *Harmo Robots Ltd* *P.O. Box 20, Burrowfield* *Welwyn Garden City, Herts, UK* *(Tel: 0438 728444)*		
Haynes & Fordham Unit 4, Moorfield Ind Est Yeadon Leeds LS19 7BM *(Tel: 0532 507090)*	*See –* *Star* *Taiyo* *Shir Miewa* *Fuji Electric*		
Head Wrightson Machine Co Ltd Durham Street Middlesborough Cleveland TS2 1PN, UK *(Tel: 0642 241155)*			
Hellstar Corporation 1600 N Chestnut Wahoo NE 680066 USA			
Hirata Industrial Machines 5-4 Myotaiji-machi Kumamoto 860 Japan	UK: *See Airstead* *Industrial Systems*	US: *Hirata Corp of America* *Indianapolis* *(Tel: 317 846 8859)*	
Hitachi Ltd Industrial Components & Equipment Div 4–1 Hammatsu-cho 2-chome Minato-ku, Tokyo, Japan *(Tel: 03/435 4272)*	UK: *GEC Industrial Robots* *Lansing Industrial Robots*	EUROPE: *Hitachi Europe GmbH* *Jaegerhofstr 32* *D–4000 Dusseldorf, FGR*	US: *Hitachi America, 6 Pearl Court* *Allendale, New Jersey 07401, USA* *(Tel: 201 825 8000)* *and General Electric Automation Systems and Automatix*

Hobart Brothers Co 600 W Main Troy OH 45373 USA	NETHERLANDS: *Hobart Bros Int'l* *Postfach 8144* *NL-1005 AC Netherlands*		
Hodges Robotics International Corp 3710 North Grand River Avenue Lansing Michigan 48906, USA *(Tel: 517 323 7427)*			
Holbronn Freres 4 rue Jeane Moulin 94130 Nogent sur Marne *(Tel: (1) 873 6945)*			
Hydro Machine Tools Ltd *(See 600 -Fanuc Robot Systems Ltd)*	*Fanuc*		

I

IBM Advanced Manufacturing Systems 1000 NW 51st Street Baco Raton Florida 33432 *(Tel: (305) 998-2000)*	UK: *IBM UK* *PO Box 41* *Portsmouth* *PO6 3AU*	WEST GERMANY: *IBM* *Advanced Manufacturing Systems* *Munich*	
IGM Industrie Zentrum Mo-sud Strasse, 2a Halle M8 A-2351 Wiener Neudorf Austria			
Ichikoh Engineering Co Ltd 1297-3 Ninomiya-cho Maebashi City Japan *(Tel: 2072/68 2131)*			
Ikegai Iron Works 1–21 Shiba 4-chome Minato-ku Tokyo, Japan 108 *(Tel: (03) 452-8111)*	EUROPE: *Ikegai Europe GmbH* *Strahlenberger Weg 8* *600 Frankfurt am Main 70, FRG* *(Tel: 0611–6160 40)*	FRANCE: *Halbronn Freres* *4 rue Jean Moulin* *94130 Nogent sur Marne* *(Tel: (1) 873 6945)*	US: *Ikegai America Corp* *2246 N. Palmer Drive STE 108* *Schaunburg, Illinois 60195* *(Tel: (312) 397 3970)*
Ikegai America Corp 2246 N. Palmer Drive STE 108, Schaunburg Illinois 60195 *(Tel: (312) 397 3970)*	*See Ikegai Iron Works Ltd* *1–21 Shiba 4-chome* *Minato-ku, Tokyo 108* *(Tel: (03) 452 8111)*		
INA Automation Ltd Forge Lane, Minworth Sutton Coldfield West Midlands B76 8AP, UK *(Tel: 021 351 4047)*			
Industria 28 avenue Clara 94420 Le Plessis Trevise France *(Tel: 01/576 53 78)*			

Industrial Automates Inc
6123 W. Mitchell St
Milwaukee
Wisconsin 53214
(Tel: (414) 327 5656)

Inser SA
José Ortega y Gasset 62
Madrid

Intarm
P.O. Box 53
Dayton, Ohio 45409
(Tel: (518) 294 0834)

Intelledex Inc
33840 Eastgate Circle
Coravillis
OR 97333
USA

International Robotmation Intelligence
2281 Las Palmas Drive
Carlsbad
Calif 92008, USA
(Tel: 714 438 4424)

ISI Manufacturing Inc
31915 Groesbeck Highway
Fraser
Michigan 48026, USA
(Tel: 313 294 9500)

J

Jobs SpA
via Marcolini 11
29100 Piacenza
Italy

UK:
Fairey Automation
Techno Trading Estate, Bramble Road
Swindon, Wilts SN2 6HB
(Tel: (0793) 481161)

Jungheinrich KG
Friedrich-Ebert-Damm 184
2000 Hamburg 70
West Germany
(Tel: 040 66 43 50)

K

Kasto-France
6 rue Pierre et Marie Curie
94430 Chennevieres sur Marne
France
(Tel: 576 20 13)

See Kitamura

Kaufeldt AB
(See Satt-Kaufeldt)

UK:
Ringway Power Systems
Churchill House, Talbot road
Old Trafford, Manchester M16 0PD
(Tel: (061) 872 6829)

Kayaba Industry Co Ltd Engineering Administration Dept 4–1 Hammatsu-cho 2-chome Minato-ku, Tokyo, Japan *(Tel: 03/435 3511)*	*See Kawasaki Heavy Industries*	
Kawasaki Heavy Industries Ltd Hydraulic Machinery Div 4–1 Hammatsu-cho 2-chome Minato-ku, Tokyo, Japan *(Tel: 03/435 6853)*	UK: *Unimation*	US: *Unimation*
Keller Technology Corporation Robotics Automation Systems 2320 Military Road Tonawanda NY 14150, USA		
Kitamura 1870 Toide Takao City Tokyo, Japan	*See Kasto France*	
Kobe Steel Ltd Machinery & Engineering Div 8–2 Marunouchi 1-chome Chiyoda-ku, Tokyo, Japan *(Tel: 03/218 7553)*	*See Trallfa*	
Komatsu Ltd 3–6 Akasaka 2-chome Minato-ku, Tokyo Japan *(Tel: 03/584 7111)*		
Koyo Automation Systems Co Ltd 62–3 Toyocho 1-chome Koto-ku, Tokyo Japan *(Tel: 03/615 2611)*	US: *Westinghouse Electric Corp* *400 Media Drive, Pittsburg* *Pennsylvania 15205* *(Tel: 412 778 4347)*	
Kremlin Spray Painting Equipment Ltd 839 Yeovil Road Slough Berks SL1 4JA, UK *(Tel: 0753 38143)*	*See AKR*	
KUKA Schweissanlagen + Roboter Zugspitzstrasse 140 D-8900 Augsburg 43 West Germany *(Tel: 0821 7971)*	UK: *KUKA Welding Systems and Robot Ltd* *7 Union Street* *Bedford, UK* *(Tel: 0234 212546)*	US: *Expert Automation Inc* *40675 Mound Road* *Stirling Heights* *Mi 48078*
Kurogane Crane Co Ltd 60 Shibacho, Minami-ku Nagoya City Japan *(Tel: 052/822 3211)*		
Kyoritsu Engineering Co Ltd Miyado Building, 6–19 Hacchobori Naka-ku Hiroshima Pref Japan *(Tel: 0822/28 9747)*		
Kyoshin Electric Co Ltd 20–7 Ikegami 6-chome Ota-ku, Tokyo Japan *(Tel: 03/751 2131)*		

L

Lamberton Robotics Ltd
26 Gartsherrie Road
Coatbridge
Strathclyde ML5 2DL, UK
(Tel: 0236 26177)

Lamson Corp
P.O. Box 4857
Syracuse
New York 13221, USA
(Tel: 315 432 5467)

Lanquepin
8 rue Proudhon
93214 La Plaine St Denis
(Tel: (1) 203 0381)

Lansing Industrial Robots — *See Hitachi*
Kingsclere Road
Basingstoke
Hants, UK
(Tel: 0256 3131)

Lincoln Electric Ltd — *See Yaskawa*
Welwyn Garden City
Herts AL7 1QA
UK
(Tel: 070732 24581)

Livernois Automation Co
25315 Kean
Dearborn
Michigan 48124
(Tel: (312) 278 0201)

Lynch Machinery Corp
2300 Crystal St
P.O. Box 2477
Anderson, Indiana 46018
(Tel: (317) 643 6671)

M

Machine Intelligence Corp — *See Yaskawa*
330 Potero Ave
Sunnyvale, ca 94086, USA
(Tel: 408 737 7960)

Mack Corp
3695 East Industrial Drive
Flagstaff
Arizona 86001, USA
(Tel: 602 526 1120)

Manca Inc — *See Fibro GmbH*
Link Drive
Rockleigh
New Jersey 07647
(Tel: (201) 767 7227)

Mantec Gmbh Postfach 2620 D-8520 Erklangen *(Tel: 09131/16200)*	*See Siemen*		
Marol Company Ltd 1.34 2-chome, Ohashi-cho Nagata-ku, Kobe *(Tel: (078) 611 2151)*	FRANCE: *CGMS* *98 rue d'Ambert, BP 1825* *45008 Orleans cedex* *(Tel: (38) 86 2514)*	UK: *Telehoist Ltd* *Manor Road* *Cheltenham* *Glos. GL51 9SH*	
Messer Griesheim Gmbh Landsbergerstrasse 432 D-8000 Munich 60 W. Germany	FRANCE: *Messer Griesheim France* *55 rue du Pelvoux, Z.I. Petite Montagne* *91019 Evry Cedex, France* *(Tel: 6/077 05 27)*		
Martonair Ltd St Margarets Road Twickenham TW1 1RJ UK *(Tel: 01 892 4411)*	WEST GERMANY: *Martinair GmbH* *4234 Alpen/Niedevrhein* *(Tel: 028 02 49 1)*	US: *Lehigh Fluid Power Inc* *PO Box 246, York Road, Route 202* *Lambertville NJ 08530* *(Tel: 609 397 3487)*	
Marwin Production Machines Waddons Brook Wednesfield Wolverhampton WV11 3AA, UK *(Tel: 0902 65363)*	*See* *Pickomatic Systems* *Robert Bosch GmbH*		
Matsushita Industrial Equipment Co Ltd Welding Apparatus Dept 1—1 Inazo-cho 3-chome Toyonaka City, Osaka, Japan *(Tel: 06/862 1121)*			
Meidensha Electric Mfg. Co Ltd Mechatronics Business Div 2–1–17 Osaki Shinagawa-ku, Tokyo, Japan *(Tel: 492/1111)*			
Microbo 3 avenue Beauregard CH 2035 Corcelles Switzerland *(Tel: 1941 3831 57 31)*	UK: *Concentric Production Research* *Upper Holland Road* *Sutton Coldfield, West Mids B72 1RD, UK* *(Tel: 021 355 1266)*		
Microbot Inc 453-H Ravendale Drive Mountain View Calif 94043, USA *(Tel: 415 968 8911)*	UK: *Sykes Instrumentation Co* *Syke House, 117–119 Station Road* *Liss, Hampshire GU33 7AJ* *(Tel: 073 082 3821)*		
Mitsubishi Heavy Industries Co Ltd Precision & Machinery Division 5–1 Marunouchi 2-chome Chiyoda-ku, Tokyo, Japan *(Tel: 03/212 3111)*	UK: *Mitsubishi Corp* *Bow Bells House, Broad St* *London EC4M 9BQ* *(Tel: 01-248 8321)*	AUSTRIA: *Voest Alpine AG* UK: *See Ringway Power Systems*	US: *Mitsubishi Corp* *277 Park Ave* *New York, NY10017, USA* *(Tel: 212 826 2188)*
Mitsubishi Electric Corporation Engineering Dept 2–3 Marunouchi 2-chome Chiyoda-ku, Tokyo, Japan *(Tel: 03/218 2722 – 2729 & 3535)*	US: *Westinghouse Electric Corp* *400 Media Drive* *Pittsburg, Penn 15205* *(Tel: 412 778 4347)*		
Mizano Iron Works Kanimachi Kanigun Gifu Pref 509 02 Japan	*See Holbronn Freres*		

Mobot Corp
980 Buenos Avenue
San Diego
Calif 92110, USA
(Tel: 714 275 4300)

Modern Machine Works Inc
Machine Tool Division
5355 S Kirkwood Avenue
Cudahv, WI 53110
USA

Modular Robotic Systems Ltd
30/31 St George's Square
Worcester WR1 1HX
UK
(Tel: 0905 612881)

Motoda Electronics Co Ltd 32–9 Kamikitazawa 4-chome Setagaya-ku, Tokyo Japan *(Tel: 03/303 8491)*	FRANCE: *CGMS* *98 rue d'Ambert BP 1825* *45008 Orleans Cedex* *(Tel: (38) 86 25 14)*

Mouldmation Ltd
Unit 1c
Chase Park Industrial Estate
Ring Road, Walsall WS7 8JQ, UK
(Tel: 05436 2075)

N

Nachi-Fujikochi Corporation Machine Tool Division, World Trade Centre 4–1 Hammatsucho, 2-chome, Minato-ku Tokyo, Japan *(Tel: 03/435 5111)*	UK: *Nachi (UK) Ltd* *70 Cardigan St* *Gosta Green, Birmingham* *(Tel: 021 359 5011)*	US: *Nachi American Inc* *223 Veterans Bld* *Carlstradt NJ 07072* *(Tel: 0201 935 8620)*

Nagoya Kiko
38 Mori Koshi, Shinden-cho
Toyoake City
Aichi Pref, Japan
(Tel: (0562) 92 7111)

Nimak – MAG Postfach 192 D-5248 Wissen West Germany *(Tel: 02742 4024/4025)*	*See* *United Technologies* *United States Robots*
Nippon Electric Co Ltd Production Facilities Development Div 1–17 Shibaura 2-chome Minato-ku, Tokyo, Japan *(Tel: 03/451 5131)*	US: *NEC America* *277 Park Avenue* *New York NY 10017* *(Tel: 212 758 1666)*

Nippon Robot Machine Co Ltd
73 Yohge Nihongi-cho
Anjoh City, Aichi Pref
Japan
(Tel: 0566/74 1101)

Nokia Robotics AB Matinkatu 22 02230 Espoo 23 Finland *(Tel: 358 0 8035610)*	US: *See Unimation*

Norda SpA via Vallecamonica 14/F 25100 Brescia Italy	FRANCE: *Dimenco PSP* *16 rue Gay Hussae* *25000 Besancon* *(Tel: 81 53 8132)*			
Nordson Corp 555 Jackson Street Amherst Ohio 44001, USA *(Tel: 216 988 9411)*	UK: *Nordson UK Ltd* *Ashurst Drive, Cheadle Heath* *Stockport SK3 0RY, UK* *(Tel: 061 491 2882)*	FRANCE: *Zl. Nord de Torcy* *77202 Marne-le Vallee Cedex 02* *(Tel: (6) 005 92 10)*	ITALY: *Basfer SpA* *via Iseo 60* *20052 Monza*	*See Yaskawa*

O

Oerlikon Buhrle AG *(See Schweissindustrie Oerlikon Buhrle AG)*	*See Volkswagen* *Hitachi*		
Oficina Tecnica Comercial (OTC) Padilla, 382,5° Barcelona Spain *(Tel: (93) 309 6462)*			
Oglaend 4301 Sandes Norway *(Tel: 04 605000)*	UK: *Fairey Automation*		
Okamura Corporation Uragocho 5-chome Yokosuka City, Kanagawa Pref. Japan *(Tel: 0468/65 8201)*	US: *C Itoh America Inc* *21415 Civic Center Drive* *Southfield, Mi 48076* *(Tel: 313 352 6250)*		
Olivetti OCN SpA Stradele Torino 10090 S Barnardo D'Ivrea (Turin) Italy	UK: *Olivetti Ltd* *105–107 Watling Street* *Bletchley, Milton Keynes, UK* *(Tel: 0908 71155)*	SPAIN: *Hispano Olivetti SA* *Gran Via 860* *Barcelona 18*	US: *Westinghouse Electric Corp* *400 Media Drive* *Pittsburg, Penn 15205* *(Tel: 412 778 4347)*
Orii Corporation 22–12 Haginaka 3-chome Ohta-ku, Tokyo Japan *(Tel: 03/744-3131)*			
Osaka Transformer Co Ltd 1–11 Tagawa 2-chome Yodogawa-ku, Osaka Japan *(Tel: 06/301 1212)*	UK: *Norman Butters Co Ltd* *PO Box 100* *Coventry CV5 6HW* *(Tel: 0203 71121)*		
Ottensener Eisenwerk GmbH Steinwerder D-2000 Hamburg 11 West Germany *(Tel: 040 306859)*			
Oy W Rosenlaw AB PB 51 SF–28101 Pori 10 Finland			

P

Pavesi International Burlington Ontario, Canada *(Tel: 416 631 6909)*			
Pendar Robotics Ltd Unit 10, Rassau Industrial Estate Ebbw Vale Gwent NP3 5SD, UK *(Tel: 0495 307070)*	US: *Pendar/Kencontronics Inc* *1821 University Avenue* *St Paul, MN 55104* *USA*	CANADA: *Can-Eng Sales* *PO Box 628* *6-800 Montrose Rd* *Niagara Falls*	AUSTRALIA: *Hawker Siddeley Brenco* *455-459 Aurburn Road* *Hawthorne* *Victoria 3122*
Pental Co Ltd 1-8 Yoshi-cho, 4-chome Soka City Santoma Pref 340 *(Tel: 0489 22:1111)*	US: *Pentel of America* *2715 Columbia St* *Torrance* *CA 90503, USA*		
Pfaff Postfach 3020/3040 D-6750 Kaiserslautern West Germany *(Tel: 0631 881)*			
Pharemme Les Nouhauts B.P. 1 87370 Saint Sulpice Lauriere France *(Tel: 55/71 44 11)*			
Pickomatic Systems Inc 37950 Commerce Drive Sterling Heights Michigan 48077, USA *(Tel: 313 939 9320)*	UK: *Marwin Production Machines* *Waddons Brook* *Wednesfield, Wolverhampton WV11 3SS* *(Tel: 0902 65363)*		
Pragma *(See DEA SpA)*			
Positech Corporation 114 Rush Lake Road Laurens IA 50554 USA			
Prab Robots Inc 5944 E Kilgore Road Kalamazoo Michigan 49003, USA *(Tel: 616 349 8761)*	EUROPE: *FN Robotics SA* *Shell Building* *60 Rue Ravenstein* *B–1000 Brussels*	JAPAN: *Murata Machinery Ltd* *To-So Building* *1–9–2 Toranomon* *Minato-Ku, Tokyo 105*	CANADA: *Can-Eng Sales* *Niagara Falls* *Ontario*
Precision Robots Inc 6 Carmel Circle Lexington Mass 02173 *(Tel: 617 862 1124)*			
Prima Progetti SpA Strada Carignano 48/2 10024 Moncalieri (Turin) Italy			

Produtec GmbH
Heilbronnerstrasse 67
D-7000
Stuttgart 1

R

Ransburg — *See Cybotech*

Rediffusion Simulation — *See American Robot Corp*
Gatwick Road
Crawley
West Sussex

Reeves Robotics Inc
Box S
Issaquah
Washington 98027, USA
(Tel: 206 392 1447)

Walter Reis GmbH & Co
D-8753 Obernburg
West Germany
(Tel: 04 188 113)

US:
Reis Machines
1426 Davis Road
Elgin, IL 60120
USA

Renault
(See Acma Cribier)

Ron-Con Ltd/Bra-Con Industries
12001 Globe Road
Livonia
MI 48154
USA

Robox Elettronica Industriale
36 via Sempione -Strada Privata Mainini
28053 Castelletto Ticino Novara
Italy
(Tel: 0331 922086)

Ringway Power Systems Ltd
Churchill House, Talbot Road
Old Trafford
Manchester M16 0PD, UK
(Tel: 061 872 6829)

See Satt-Kaufeldt
Mitsubishi Heavy Industries

Robotic Sciences International Inc
2709 South Halladay
Santa Ana
Calif 92705, USA
(Tel: 714 979 6831)

Robotiks Inc
507 Prudential Road
Horsham
Pennsylvannia 19044, USA
(Tel: 215 674 2800)

S

Sandhu Rhino Robots
308 S State Street
Champaign
Illinois 61820, USA
(Tel: 217 352 8485)

Sanki Engineering Co Ltd
Industrial Plant & Machinery Div
Sales No 1 Dept, Sanshin Building,
4–1 Yurakucho 1-chome Chiyoda-ku, Tokyo, Japan
(Tel: 03/502 6111)

Sankyo Seiki Mfg. Co Ltd Machine Tools Division 17–2 Shinbashi 1-chome Minato-ku, Tokyo, JAPAN *(Tel: 03/591 6886)*	UK: *IBM UK* *PO Box 41* *Portsmouth PO6 3AU*	FRANCE: *CGMS* *98 rue d'Ambert BP 1825* *45008 Orleans Cedex* *(Tel: (38) 86 25 14)*	US: *IBM* *1000 NW 51st Street* *Baco Raton, Fl 33432* *(Tel: 305 998 2000)*
Satt-Kaufeldt AB P.O. Box 32 006 S–12611 Stockholm Sweden *(Tel: 08 810100)*	UK: *Ringway Power Systems* *Churchill House, Talbot Road* *Old Trafford, Manchester M16 OPD* *(Tel: (061) 872 6829)*		
SCEMI 61 rue de Funas 38300 Bourgoin-Jallieu France *(Tel: 74/93 20 04)*	*See Yaskawa*		
Schrader-Bellows US Rt. 1 N Wake Forest North Carolina 27587, USA *(Tel: 919 556 4031)*	UK: *Schrader-Bellows* *Bridgetown* *Cannock, Staffs* *(Tel: 05435 2644)*		

Schweissindustrie Oerlikon Buhrle AG
Birchstrasse 230
8050 Zurich
Switzerland
(Tel: 01 301 2121)

Sciaky 119 quai Jules Guesde 94400 Vitry/Seine France *(Tel: 680 85 07)*	UK: *Sciaky Electric Welding Machines Ltd* *212 Bedford Avenue, Slough* *Berks SL1 4RH, UK* *(Tel: 0753 25551)*	BELGIUM: *AIMO* *64–68 rue Lesbroussart* *B1050* *Brussels*	AUSTRIA: *Otto Dohmen & Co* *Argentinienstrasse 42* *A1041 Vienna*
	GERMANY: *M. Nitsche* *Ottilienstrasse 7a* *D8000 Munich*	ITALY: *Ravedati Eng* *Corso Montevecchio 46* *10129 Turin*	SWEDEN: *Sciaky Scardia* *Skäpplandsgatan 4* *S 41478, Gothenburg*
	EAST GERMANY: *Sciaky* *Sleinmetzstrasse 2* *D 62 Weisbaden*	SPAIN: *Sciaky* *Castaner 23* *Barcelona 22*	
Seiko Instruments USA Inc 2990 W. Lomita Boulevard Torrance Calif 90505, USA *(Tel: 213 530 8777)*	JAPAN (parent company): *Daini Seikosha Co Ltd* *31-1,6-chome* *Kameido, Koto-ku, Tokyo 136* *(Tel: (03) 682 1111)*	ITALY: *Jocam Sri* *via Polidoro da Caravoggio 37* *20156 Milan*	UK: *See Airstead Industrial Systems*

Sietam 38–48 avenue du President Kennedy 91170 Vitry-Chatillon France *(Tel: 6/996 91 80)*			
Shinko Electric Co Ltd 3–12-2 Nihonbashi Chuo-ku, Tokyo Japan *(Tel: 03/222 2721)*			
Shinmeiwa Industry Co Ltd Machinery & Plant Division 1–1 Shinmeiwa-cho Takarazuka City, Hyogo Pref., Japan *(Tel: 0798/52 1234)*	UK: *Haynes & Fordham Ltd*	FRANCE: *Commercy* *55200 Commercy* *(Tel: (29) 01 01 04)*	US: *Nichimen Colne* *6 North Michigan Ave* *Chicago, Ill 60602, USA* *(Tel: 312 346 9339)*
Shoku Corporation Engineering Department 1010 Minorudai, Matsudo City Chiba Pref., Japan *(Tel: 0473/64 1211)*	WEST GERMANY: *VFW Fokker GmbH* *Werk Haykemkamp* *2870 Delmenhorst* *(Tel: 04221 131)*		
Siemens AG Rupert-Mayer-Strasse 44 8000 Munchen 70, Postfach 70 00 75 West Germany *(Tel: 089 722 26126)*	*See* *Fanuc* *Mantec*		
Sigma Sales Inc 6505C Serrano Avenue Anaheim Hills Calif 92807, USA *(Tel: 714 974 0166)*			
Sodimat Sapignies 62121 A 62121 Achiet-le Grand, Cedex 6 France *(Tel: 6/903 78 79)*	SWITZERLAND: *Walther* *Les Breneb* *CH 2416*		
Shawa Kuatuski 3–19 Kanda-Sakumacho Chiyada-ku, Tokyo Japan	EUROPE: *VFW Fokker GmbH* *Werk Haykemkamp* *2870 Delmenhurst* *(Tel: 04221 131)*		
600 Fanuc Robotics Ltd Hythe Station Road Colchester Essex, UK *(Tel: 0206 870380)*	*See Fanuc*		
Sormel SA rue Becquerel Z.I. Chateaufarine, B.P. 1565 25009 Besancon Cedex, France *(Tel: 8181 4245)*	US: *Ceeris International Inc* *1055 Thomas Jefferson St N.W. Ste 414* *Washington DC 20007* *(Tel: (202) 342 5400)*	US: *Sormel/Black & Webster* *281 Winter Street* *Waltham, Mass 02254* *(Tel: (617) 890 9100)*	
Spine Robotics AB Flojelbergsgaten 14 S 43137 Molindal Sweden *(Tel: 031 870710)*			
Star Seiki Co Ltd 252 Kawachiya Shinden Komaki City Japan *(Tel: 0568/75 5211)*	UK: *Haynes & Fordham Ltd*	WEST GERMANY: *DME Deutschland* *D 7106 Neuenstadt* *(Tel: 07139 511)*	US: *Sterltech* *PO Box 23421* *Milwaukee, Wi 53223* *(Tel: 414 354 0493)*

Sterling Detroit Company
261 E. Goldengate
Detroit
Michigan, 48203, USA
(Tel: 313 366 3500)

Sterltech
PO Box 23421
Milwaukee, Wi 53223
(Tel: 414 354 0493)

Sumitomo Heavy Industry Ltd
Shumisho Building, 1 Mitoshirocho
Chiyoda-ku, Tokyo
Japan
(Tel: 03/296 5183)

Systems Control Ltd
Perry Avenue
Teeside Industrial Estate, Thornaby
Cleveland TS17 9LN, UK
(Tel: 0642 764823)

US:
TecQuipment Inc
PO Box 1024
Acton, Mass 01820
(Tel: (617) 263 1767)

T

Taiyo Ltd
48 Kitaguchi-cho
Higashiyodogawa-ku
Osaka, Japan
(Tel: 06/329 1111)

UK:
Haynes & Fordham Ltd

FRANCE:
CGMS
98 rue d'Ambert BP 1825
45008 Orleans Cedex
(Tel: (38) 86 25 14)

Taylor Hitec Ltd
77 Lyons Lane
Chorley, Lancs PR6 0PB
(Tel: 02572 65825)

TecQuipment Inc
PO Box 1074, Acton
Mass 01720
(Tel: (617) 263 1767

See Systems Control Ltd

Thermwood Corporation
P.O. Box 436
Dale
Indiana 47523, USA
(Tel: 812 937 4476)

See Binks Bullows (US)
Cyclomatic
Didde Graphics

Tokico Ltd
Hydraulic & Pneumatic Equipments Sales Dept
6–10 Uchikanda 1-chome
Chiyoda-ku, Tokyo, Japan
(Tel: 03/292 8111)

USA:
Tokico America Inc
3888 W Lomita Boulevard/Suite E
Torrance
CA 90505, USA

Tokyo Keiki Co Ltd
2–16 Minami Kamata
Ohta-ku, Tokyo
Japan
(Tel: 03/732 2111)

Torsteknik AB
Box 130
S–385 00 Torsas
Sweden

UK:
Torsteknik AB (UK)
3 West Bar
Banbury
Oxon OX16 9SD

See Yaskawa

Toshiba Seiki Co Ltd Sales Division 14–33 Higashi Kashiwagaya 5-chome Ebina City, Kanagawa Pref., Japan *(Tel: 0462/31 8111)*	UK: *Evershed Robotics Ltd* *Bridge Road* *Chertsey* *Surrey KT16 8LJ*	FRANCE: *CGMS* *98 rue d'Ambert BP 1825* *45008 Orleans Cedex* *(Tel: (38) 86 25 14)*	US: *Fleximation Systems Corp* *53 Second Ave, Burlington* *Mass 01803* *(Tel: (617) 229 6670)*

Toyoda Machine Works Ltd
1 Asahi-cho, 1-chome
Katiya City Aichi Pref.
Japan
(Tel: 0566/22 2211)

Trallfa P.O. Box 113 N 4341 Bryne Norway *(Tel: 04 48 18 00)*	UK: *DeVilbiss Co Ltd* *Ringwood Road* *Bournemouth, Hants* *(Tel: (0202) 571111)*	FRANCE: *Toussaint-DeVilbiss* *163–171 avenue des Aureats* *26014 Valence Cedex* *(Tel: (75) 44 79 77)*	US: *The DeVilbiss Company* *300 Phillips Ave, PO Box 913* *Toledo, Ohio 43692* *(Tel: (419) 470 2169)*	WEST GERMANY: *Trallfa Gmbh* *Frankfurter-strasse* *D-6078 Neu-Isenburg* *(Tel: 0610 2 37627)*	SPAIN: *Harry Walker* *Techno-quip* *Barcelona 15* *(Tel: 243 3107)*

U

Unimation Inc Shelter Rock Lane Danbury Connecticut, 06810, USA *(Tel: 203 744 1800)*	UK: *Unimation (Europe) Ltd* *Unit C, Stafford Park 18* *Telford, Salop, UK* *(Tel: 0952 618931)*	FINLAND: *Nokia Automation* *Matinkau 22* *02230 Espoo 23*	SPAIN: *Inser SA* *Madrid*	JAPAN: *Kawasaki Heavy Industries* *4–1 Shiba Hamamstu-cho-2-chome* *Minato-ku, Tokyo* *(Tel: (03) 435 6907)*

Union Carbide Deutschland GmbH
(See also Nimak – MAG)
Postfach 133, D–5248 Wissen
West Germany
(Tel: 02742 751)

United States Robots Inc — *See Nimak*
1000 Conshohocken Road
Conshohocken
PA 19428
USA

United Technologies Automotive Group — *See Nimak*
5200 Auto Club Drive
Dearborn
MI 48126
USA

V

VFW Fokker GmbH — *See Shoku Corp*
Showa Kuatuski
Hunefeldstrasse 1–5
D-2800 Bremen 1
W. Germany

VSI Automation Assembly Inc
165 Park Street
Troy
MI 48084
USA

Voest Alpine AG — *See*
Ringway Power Systems
Mitsubishi Heavy Industries
Postfach 2
A 4010 Linz
Austria
(Tel: 0732 585 1)

Volkswagen AG Industrial Robot Division Wolfsburg West Germany	USA: *General Electric Corp*		

W

Westinghouse Electric Corporation *See Unimation Inc*
400 Media Drive
Pittsburg
Pennsylvania 15205, USA
(Tel: 412 778 4347)

Wexford Robotics Ltd
2118 Queen
Regina Sask
Canada S4T4C3
(Tel: (306) 522 7469

Wickman Automated Assembly Ltd
Herald Way, Brandon Road
Binley
Coventry CV3 2NY, UK
(Tel: 0203 45080)

WRA Ltd Units 2/3, Wulfrun Trading Estate Stafford Road Wolverhampton, UK *(Tel: 0902 711201)*	*See Carl Cloos GmbH* *IGM (Limat) Austria* *Unimation*

Y

Yaskawa & Co Ltd Robot, Material Handling Division 16–9 Uchikanda 2-chome Chiyoda-ku Tokyo, Japan *(Tel: 03/256 72711)*	UK: *Lincoln Electric Ltd* *Welwyn Garden City* *Herts AL7 1QA* *(Tel: 070732 4581)*	FRANCE: *SCEMI* *61 Rue de Funas* *38300 Bourgoin-Jallieu* *(Tel: (74) 93 20 04)*	US: *Yaskawa Electric America Inc* *305 Era Drive, Northbrook* *Illinois 60062, USA* *(Tel: 312 564 0770)*
	US: *Machine Intelligence Corp* *330 Potero Ave* *Sunnyvale, Ca 94086* *(Tel: 408 737 7960)*	SWEDEN: *Torsteknik* *Box 130* *S 38500 Torsa*	US: *Bendix Corp* *Nordson Corp*

Yasui Sangyo Co Ltd
3711 Mannohara-Shinden
Fujinomiya-Shi
Shizuoka-ken, Japan
(Tel: 05442 62124)

Z

ZF Postfach 25 20 D–7990 Friedrichshafen 1 West Germany *(Tel: 07541 701–1)*	UK: *Hahn & Kolb UK* *6 Forum Drive* *Leicester Road* *Rugby CV21 1NY*

Index to Advertisers

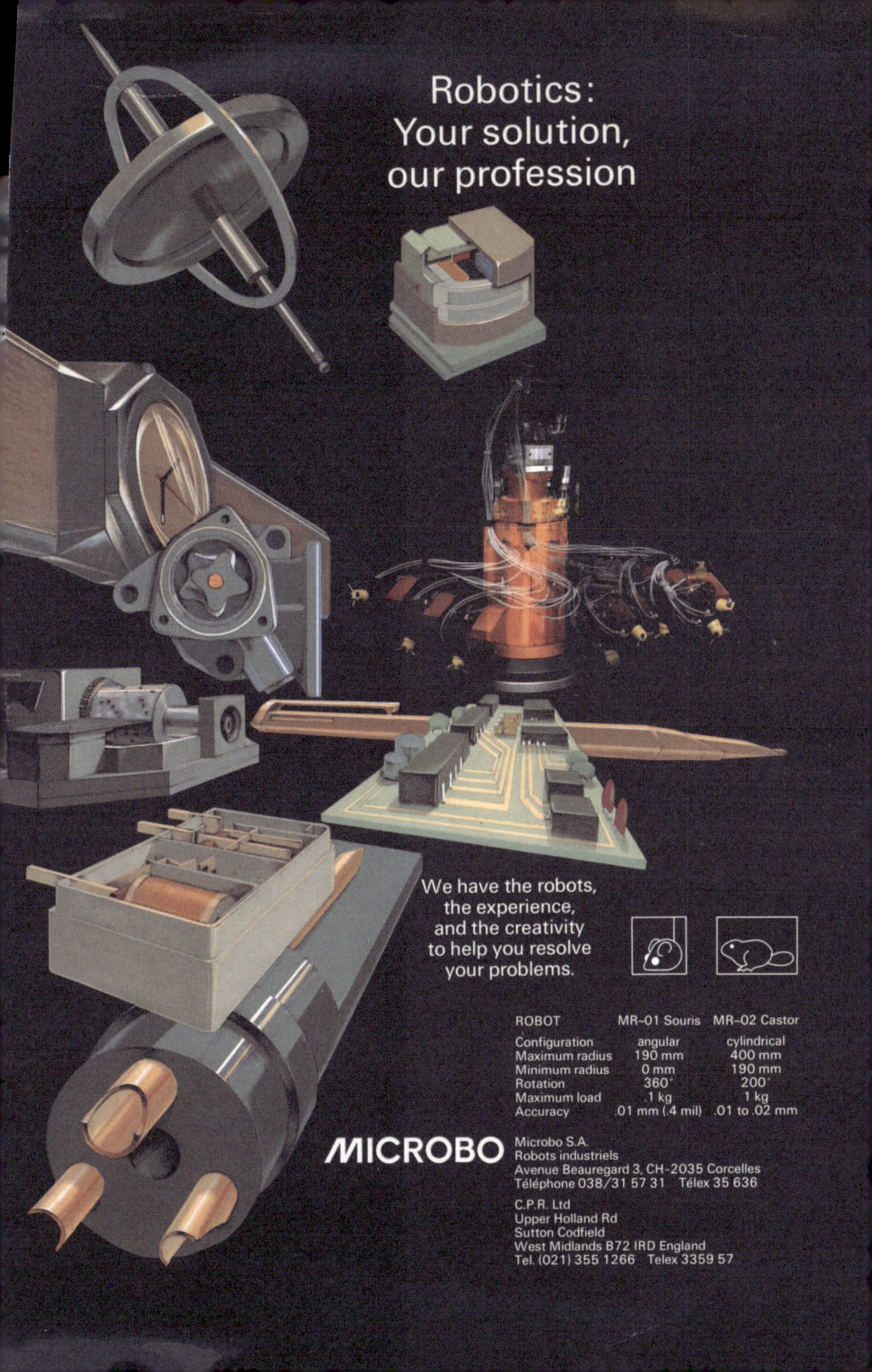
Robotics:
Your solution,
our profession
We have the robots,
the experience,
and the creativity
to help you resolve
your problems.
ROBOT MR-01 Souris MR-02 Castor
Configuration angular cylindrical
Maximum radius 190 mm 400 mm
Minimum radius 0 mm 190 mm
Rotation 360° 200°
Maximum load .1 kg 1 kg
Accuracy .01 mm (.4 mil) .01 to .02 mm
MICROBO
Microbo S.A.
Robots industriels
Avenue Beauregard 3, CH-2035 Corcelles
Téléphone 038/31 57 31 Télex 35 636
C.P.R. Ltd
Upper Holland Rd
Sutton Codfield
West Midlands B72 IRD England
Tel. (021) 355 1266 Telex 3359 57

GPSR Compliance

The European Union's (EU) General Product Safety Regulation (GPSR) is a set of rules that requires consumer products to be safe and our obligations to ensure this.

If you have any concerns about our products, you can contact us on ProductSafety@springernature.com

In case Publisher is established outside the EU, the EU authorized representative is:

Springer Nature Customer Service Center GmbH
Europaplatz 3
69115 Heidelberg, Germany

Batch number: 10845484

Printed by Printforce, the Netherlands